SPORTY'S®

WHAT YOU SHOULD KNOW™ SERIES

PTS STUDY GUIDE

Instrument Rating Practical Test Standards

for

Airplane

Cross-Referenced

to

Sporty's Interactive DVD Course

Sporty's Academy, Inc.
Clermont County/Sporty's Airport
Batavia, OH 45103

Printed in the United States of America
ISBN 978-0-9715631-7-9

For additional copies, reorder #M292A

Call: 1 (USA) 800.SPORTYS
(776.7897)

Fax: 1 (USA) 800.359.7794
1 (USA) 513.735.9200
sportys.com

02/06

(Intentionally Left Blank)

Table of Contents

Preface

Sporty's *What You Should Know*™ Complete Flight Training DVD course for the Instrument Rating has been designed to completely prepare you for instrument flying. This series is an effective tool in understanding instrument flying, and can be used as self-study for pilots working on their instrument rating or by previously rated pilots as a refresher course.

The subject matter is presented in a logical sequence that parallels the flight instruction you will be receiving. This sequence is also the best way to prepare for the FAA computerized knowledge exam. This book is not a substitute for the DVDs, but a supplement to help you completely prepare for your knowledge test, oral and practical exams, and to become a better pilot.

This study guide is arranged into two major sections.

The first section contains the Instrument Rating Practical Test Standards for Airplane with a DVD cross-reference. This section is intended to be used as a review prior to your oral and practical exams. It also may be used as a supplemental index to the DVDs. It relates the various elements of the PTS to the appropriate Sporty's DVD volumes and segments for further review.

The second section contains supplemental material that you should study after watching each DVD volume. This information will support the subjects presented by the related DVDs and will provide reinforcing notes or may be used as a quick reference. Some of these subsections have an additional group of illustrations applicable to the instructional material on that DVD and/or FAA test questions found on the DVD.

Most DVD video segments conclude with a set of optional interactive FAA test questions, answers, and explanations. Some of the questions refer to illustrations that are included on the DVDs. The more complex illustrations are also reproduced in this book, after the notes for that volume.

There are also additional approach charts in this book that are not part of any test question, but are included to help you follow along with the VOR and NDB segments of Volume 3, and the GPS approaches in Volume 6.

This study guide *is not* intended to stand alone. It is a part of the total training package supplied with Sporty's *What You Should Know* Complete Flight Training DVD course for the Instrument Rating.

Maximum benefit can be derived from this course by following the instructions below:

> After viewing each DVD segment, answer any FAA test questions that follow. The answers and explanations are provided with each question.
>
> After finishing a Sporty's DVD Volume, read the notes subsection for that particular volume. The notes will reinforce key concepts from the DVDs through illustrations and explanations of these points.
>
> Take the review test for that volume (the last item on the DVD main menu). This test combines all of the quiz questions from each segment and provides you with a score for gauging your progress.
>
> When you have finished the review test for a DVD volume, go on to the next volume, or review the video if you need more exposure to certain areas. Repetition of this process can greatly enhance your ability to understand difficult topics.

Best of luck with your studies and welcome to your new adventure.

Sporty's Academy, Staff
February, 2006
Batavia, Ohio

Conventions Used in This Manual

The Instrument Rating Practical Test Standards (PTS) with DVD Cross-Reference contains the text of the PTS with references to information that may be found in Sporty's *Complete* Flight Training Course for the Instrument Rating on DVD for each element. The cross-reference will appear in the following format:

- A number indicating the DVD volume will be followed by a period and number indicating the segment within the DVD. For example, 3.1 would indicate to refer to Segment 1 of DVD Volume 3 from the course.

Sporty's *Complete* Flight Training Course utilizes the building block method of learning. This method assumes that you already have the knowledge of a Private Pilot and does not re-teach certain Private Pilot basics. Private Pilot knowledge elements that are evaluated in the Instrument PTS are referenced to Sporty's *Complete* Flight Training Course for the Private Pilot on DVD. The references to these DVDs are proceeded with the letters "Pvt".

Appendices and pages within this study guide and the AFM/POH for your airplane are also referenced.

FAA References Used in This Manual

Many of the references below were used by the FAA in preparing the PTS and in the preparation of this manual. Most of the references listed are books and may be purchased from Sporty's by calling 1.800.SPORTYS (776.7897) from the USA or by logging on to http://www.sportys.com.

14 CFR Part 43 Maintenance, Preventive Maintenance, Rebuilding, and Alteration
14 CFR Part 61 Certification: Pilots and Flight Instructors
14 CFR Part 91 General Operating and Flight Rules
14 CFR Part 97 Standard Instrument Approach Procedures
NTSB Part 830 Notification and Reporting of Aircraft Accidents and Incidents
FAA-H-8083-1 Aircraft Weight and Balance Handbook
FAA-H-8083-3 Airplane Flying Handbook
FAA-H-8083-15 Instrument Flying Handbook
FAA-H-8083-25 Pilot's Handbook of Aeronautical Knowledge
AC 00-2 Advisory Circular Checklist
AC 00-6 Aviation Weather
AC 00-45 Aviation Weather Services
AC 00-54 Pilot Wind Shear Guide
AC 60-28 English Language Skill Standards Required by 14 CFR parts 61, 63, and 65
AC 61-65 Certification: Pilots and Flight Instructors
AC 61-67 Stall Spin Awareness Training
AC 61-84 Role of Preflight Preparation
AC 67-2 Medical Handbook for Pilots
AC 90-45 Approval of Area Navigation Systems for Use in the U.S. National Airspace System
AC 90-48 Pilots' Role in Collision Avoidance
AC 90-94 Guidelines for Using Global Positioning Systems
AC 91-13 Cold Weather Operation of Aircraft
AC 91-43 Unreliable Airspeed Indications
AC 91-55 Reduction of Electrical Systems Failure Following Engine Starting
AC 120-51 Crew Resource Management Training
AIM Aeronautical Information Manual
A/FD Airport/Facility Directory
NOTAMs Notices to Airmen
AFM/POH - FAA-Approved Flight Manual/Pilot Operating Handbook
Approach Charts and Legends
En Route Charts and Legends
DP and STAR Charts and Legends

(Intentionally Left Blank)

Section 1 - Instrument Rating Practical Test Standards - Airplane with DVD Cross-Reference

I. AREA OF OPERATION: PREFLIGHT PREPARATION

A. TASK: WEATHER INFORMATION

NOTE: Where current weather reports, forecasts, or other pertinent information is not available, this information will be simulated by the examiner in a manner that will adequately measure the applicant's competence.

DVD Volume.Segment

Objective. To determine that the applicant:

1. Exhibits adequate knowledge of the elements related to aviation weather information by obtaining, reading, and analyzing the applicable items, such as- ....4.4, 4.5, 5.2, 5.3, 5.6, 5.10, 5.11, 5.12, 5.13, 7.4, Pvt.3.10, Pvt.3.11, Pvt.5.13, Pvt.6.12, PTS Study Guide Page 2-36, PTS Study Guide Page 2-56
   a) weather reports and forecasts....4.4, 5.2, 5.11, 7.4, Pvt.3.11, Pvt.5.13, Pvt.6.12
   b) pilot and radar reports....4.4, 5.3, 5.12, Pvt.5.13, Pvt.6.12
   c) surface analysis charts....5.2
   d) radar summary charts....4.4, 4.5
   e) significant weather prognostics....7.4, Pvt.5.13
   f) winds and temperatures aloft....4.4, 5.3, 5.13, 7.4, Pvt.6.12
   g) freezing level charts....Appendix A
   h) stability charts....Appendix A
   i) severe weather outlook charts. (Now called Convective Outlook--*Ed.*)....7.4, Pvt.6.12
   j) SIGMETs and AIRMETs....4.4, 5.6, Pvt.6.12
   k) ATIS reports....7.14
2. Correctly analyzes the assembled weather information pertaining to the proposed route of flight and destination airport, and determines whether an alternate airport is required, and, if required, whether the selected alternate airport meets the regulatory requirement....4.13, 4.14, 5.10, PTS Study Guide Page 2-56

B. TASK: CROSS-COUNTRY FLIGHT PLANNING

Objective. To determine that the applicant:

1. Exhibits adequate knowledge of the elements by presenting and explaining a preplanned cross-country flight, as previously assigned by the examiner (preplanning is at examiner's discretion). It should be planned using real time weather and conform to the regulatory requirements for instrument flight rules within the airspace in which the flight will be conducted....1.3, 4.2, 4.3
2. Exhibits adequate knowledge of the aircraft's performance capabilities by calculating the estimated time en route and total fuel requirement based upon factors, such as-....4.6, 4.13, Pvt.5.5
   a) power settings....1.3, 6.7, Pvt.5.5
   b) operating altitude or flight level....1.3, Pvt.5.5
   c) wind....1.3, Pvt.5.5
   d) fuel reserve requirements....4.6, 4.13
3. Selects and correctly interprets the current and applicable en route charts, instrument departure procedures (DPs), RNAV, STAR, and Standard Instrument Approach Procedure Charts (IAP)....2.6, 3.3, 4.2, 4.3, 4.8, 4.10
4. Obtains and correctly interprets applicable NOTAM information....7.12, Pvt.5.3
5. Determines the calculated performance is within the aircraft's capability and operating limitations....AFM/POH
6. Completes and files a flight plan in a manner that accurately reflects the conditions of the proposed flight. (Does not have to be filed with ATC.)....2.11, 7.12, PTS Study Guide Page 2-5
7. Demonstrates adequate knowledge of GPS and RAIM capability, when aircraft is so equipped....6.13

II. AREA OF OPERATION: PREFLIGHT PROCEDURES

A. TASK: AIRCRAFT SYSTEMS RELATED TO IFR OPERATIONS

DVD Volume.Segment

Objective. To determine that the applicant exhibits adequate knowledge of the elements related to applicable aircraft anti-icing/deicing system(s) and their operating methods to include:

1. Airframe. 5.5
2. Propeller 5.5
3. Intake Pvt.1.11
4. Fuel. .... AFM/POH
5. Pitot-static. .... AFM/POH

B. TASK: AIRCRAFT FLIGHT INSTRUMENTS AND NAVIGATION EQUIPMENT

Objective. To determine that the applicant:

1. Exhibits adequate knowledge of the elements related to applicable aircraft flight instrument system(s) and their operating characteristics to include-
   - a) pitot-static. 7.10, 7.11, Pvt.3.13
   - b) altimeter. 1.5, 7.10, 7.11, Pvt.3.7, Pvt.3.13
   - c) airspeed indicator. 1.5, 7.10, 7.11, Pvt.3.13
   - d) vertical speed indicator. 1.5, 7.10, 7.11, Pvt.3.13
   - e) attitude indicator. 1.4, 1.5, 7.10, 7.11
   - f) horizontal situation indicator. .... 6.3, PTS Study Guide Page 2-59
   - g) magnetic compass. 1.11, Pvt.6.3
   - h) turn-and-slip indicator/turn coordinator. 1.5, 1.9, 1.11, 7.10, 7.11
   - i) heading indicator. 1.5, 7.10, 7.11
   - j) electrical systems. Pvt.1.10
   - k) vacuum systems. 1.13, 7.11
   - l) electronic flight instrument display 6.3
2. Exhibits adequate knowledge of the applicable aircraft navigation system(s) and their operating characteristics to include-
   - a) VOR. 2.3, 2.5, 3.9, 6.3, Pvt.5.7
   - b) DME. .... Pvt.5.7, PTS Study Guide Page 2-59
   - c) ILS. 3.4, 3.5, 6.3
   - d) marker beacon receiver/indicators. 3.3, 3.5
   - e) transponder/altitude encoding. 2.13
   - f) ADF. 3.11, Pvt.5.9
   - g) GPS .... 2.5, 6.13, Pvt.5.8, PTS Study Guide Page 2-59
   - h) FMS .... AFM/POH

C. TASK: INSTRUMENT COCKPIT CHECK

Objective. To determine that the applicant:

1. Exhibits adequate knowledge of the elements related to preflighting instruments, avionics, and navigation equipment cockpit check by explaining the reasons for the check and how to detect possible defects. .... 7.5, 7.7, 7.10, Appendix B
2. Performs the preflight on instruments, avionics, and navigation equipment cockpit check by following the checklist appropriate to the aircraft flown. .... 4.11, Appendix B
3. Determines that the aircraft is in condition for safe instrument flight including-
   - a) communications equipment. .... 7.7, Appendix B
   - b) navigation equipment, as appropriate to the aircraft flown: .... 6.13, 7.7, Appendix B
   - c) magnetic compass. .... Appendix B
   - d) heading indicator. .... 7.7, 7.10, Appendix B
   - e) attitude indicator. .... 7.7, 7.10, Appendix B
   - f) altimeter. .... 7.7, 7.10, Appendix B
   - g) turn-and-slip indicator/turn coordinator. .... 7.7, 7.10, Appendix B
   - h) vertical speed indicator. .... 7.10, Appendix B
   - i) airspeed indicator. .... Appendix B
   - j) clock. .... 7.7, Appendix B
   - k) power source for gyro-instruments. .... 1.13, 7.7, 7.10, Appendix B
   - l) pitot heat. .... Appendix B
   - m) electronic flight instrument display .... AFM/POH
   - n) traffic awareness/warning/avoidance system. .... AFM/POH
   - o) terrain awareness/warning/alert system. .... AFM/POH
   - p) FMS. .... AFM/POH
   - q) auto pilot. .... AFM/POH
4. Notes any discrepancies and determines whether the aircraft is safe for instrument flight or requires maintenance. .... Appendix B

III. AREA OF OPERATION: AIR TRAFFIC CONTROL CLEARANCES AND PROCEDURES

NOTE: The ATC clearance may be an actual or simulated ATC clearance based upon the flight plan.

A. TASK: AIR TRAFFIC CONTROL CLEARANCES

Objective. To determine that the applicant:

DVD Volume.Segment

1. Exhibits adequate knowledge of the elements related to ATC clearances and pilot/controller responsibilities to include tower en route control and clearance void times. 2.8, 2.12, 7.12, 7.14
2. Copies correctly, in a timely manner, the ATC clearance as issued. 2.8, 7.12
3. Determines that it is possible to comply with ATC clearance. 7.14
4. Interprets correctly the ATC clearance received and, when necessary, requests clarification, verification, or change. 2.8, 7.14
5. Reads back correctly, in a timely manner, the ATC clearance in the sequence received. 2.8, 7.14
6. Uses standard phraseology when reading back clearance. 2.8, 7.12
7. Sets the appropriate communication and navigation systems and transponder codes in compliance with the ATC clearance.

B. TASK: COMPLIANCE WITH DEPARTURE, EN ROUTE, AND ARRIVAL PROCEDURES AND CLEARANCES

Objective. To determine that the applicant:

1. Exhibits adequate knowledge of the elements related to ATS routes, and related pilot/controller responsibilities. .... 2.6, 2.12, 4.2, 4.3, 4.8, 4.10, PTS Study Guide Page 2-35
2. Uses the current and appropriate navigation publications for the proposed flight. 2.10, 4.3, 4.8, 4.10
3. Selects and uses the appropriate communication facilities; selects and identifies the navigation aids associated with the proposed flight.
4. Performs the appropriate aircraft checklist items relative to the phase of flight. 4.11
5. Establishes two-way communications with the proper controlling agency, using proper phraseology. 2.8
6. Complies, in a timely manner, with all ATC instructions and airspace restrictions. 2.6, 7.14
7. Exhibits adequate knowledge of communication failure procedures. 7.8, 7.9
8. Intercepts, in a timely manner, all courses, radials, and bearings appropriate to the procedure, route, or clearance. 2.6
9. Maintains the applicable airspeed within +/-10 knots; headings within +/-10°; altitude within +/-100 feet; and tracks a course, radial, or bearing within ¾ scale deflection of the CDI. 1.3

C. TASK: HOLDING PROCEDURES

NOTE: Any reference to DME will be disregarded if the aircraft is not so equipped.

Objective. To determine that the applicant:

1. Exhibits adequate knowledge of the elements related to holding procedures. .... 3.14, 7.8, Appendix C, PTS Study Guide Page 2-13
2. Changes to the holding airspeed appropriate for the altitude or aircraft when 3 minutes or less from, but prior to arriving at, the holding fix.
3. Explains and uses an entry procedure that ensures the aircraft remains within the holding pattern airspace for a standard, nonstandard, published, or nonpublished holding pattern. 3.14
4. Recognizes arrival at the holding fix and initiates prompt entry into the holding pattern. .... Appendix C
5. Complies with ATC reporting requirements. 2.8
6. Uses the proper timing criteria, where applicable, as required by altitude or ATC instructions. .... 3.7, 3.14, 7.8, Appendix C
7. Complies with pattern leg lengths when a DME distance is specified. .... Appendix C
8. Uses proper wind correction procedures to maintain the desired pattern and to arrive over the fix as close as possible to a specified time. .... Appendix C
9. Maintains the airspeed within +/-10 knots; altitude within +/-100 feet; headings within +/-10°; and tracks a selected course, radial, or bearing within ¾ scale deflection of the CDI. 1.3

IV. AREA OF OPERATION: FLIGHT BY REFERENCE TO INSTRUMENTS

A. TASK: BASIC INSTRUMENT FLIGHT MANEUVERS (IA, IH, PL, AA, HA, PLA, PC)

DVD Volume.Segment

Objective. To determine that the applicant can perform basic flight maneuvers.

1. Exhibits adequate knowledge of the elements related to attitude instrument flying during straight-and-level, climbs, turns, and descents while conducting various instrument flight procedures.................. 1.4, 1.5, Appendix D
2. Maintains altitude within +/- 100 feet during level flight, headings within +/- 10°, airspeed within +/- 10 knots, and bank angles within +/- 5° during turns. 1.3, 1.5
3. Uses proper instrument crosscheck and interpretation, and apply the appropriate pitch, bank, power, and trim corrections when applicable. .... 1.4, 1.5, Appendix D, PTS Study Guide Page 2-2

B. TASK: RECOVERY FROM UNUSUAL FLIGHT ATTITUDES

NOTE: Any intervention by the examiner to prevent the aircraft from exceeding any operating limitations, or entering an unsafe flight condition, shall be disqualifying.

Objective. To determine that the applicant:

1. Exhibits adequate knowledge of the elements relating to attitude instrument flying during recovery from unusual flight attitudes (both nose-high and nose-low). 7.11
2. Uses proper instrument crosscheck and interpretation, and applies the appropriate pitch, bank, and power corrections in the correct sequence to return the aircraft to a stabilized level flight attitude. 7.11

V. AREA OF OPERATION: NAVIGATION SYSTEMS

A. TASK: INTERCEPTING AND TRACKING NAVIGATIONAL SYSTEMS AND DME ARCS

NOTE: Any reference to DME arcs, ADF, or GPS shall be disregarded if the aircraft is not equipped with these specified navigational systems.

DVD Volume.Segment

Objective. To determine that the applicant:

1. Exhibits adequate knowledge of the elements related to intercepting and tracking navigational systems and DME arcs. 3.9, 3.11, 3.13, 3.18, 6.3, 6.13, Pvt.5.7
2. Tunes and correctly identifies the navigation facility. 1.3, 2.3, 3.4, Pvt.5.7
3. Sets and correctly orients the course to be intercepted into the course selector or correctly identifies the course on the RMI. 3.9, 6.3, Pvt.5.7
4. Intercepts the specified course at a predetermined angle, inbound or outbound from a navigational facility. 1.3, 3.9, 6.3
5. Maintains the airspeed within +/-10 knots, altitude within +/-100 feet, and selected headings within +/-5°. 1.3
6. Applies proper correction to maintain a course, allowing no more than three-quarter-scale deflection of the CDI or within +/-10° in case of an RMI. 1.3, 3.4, 3.9, 6.3
7. Determines the aircraft position relative to the navigational facility or from a waypoint in the case of GPS. 6.13, Pvt.5.7, Pvt.5.9
8. Intercepts a DME arc and maintain that arc within +/-1 nautical mile. 1.3, 3.18
9. Recognizes navigational receiver or facility failure, and when required, reports the failure to ATC. 2.5, 2.8, 6.13, Pvt.5.7

VI. AREA OF OPERATION: INSTRUMENT APPROACH PROCEDURES

NOTE: TASK D, Circling Approach, is applicable only to the airplane category.

NOTE: The requirements for conducting a GPS approach for the purpose of this test are explained on PTS Study Guide Page 1-13 of the Supplemental PTS Information.

A. TASK: NONPRECISION INSTRUMENT APPROACH (NPA)

NOTE: The applicant must accomplish at least two nonprecision approaches (one of which must include a procedure turn or, in the case of an RNAV approach, a Terminal Arrival Area (TAA) procedure) in simulated or actual weather conditions. At least one nonprecision approach must be flown without the use of autopilot and without the assistance of radar vectors. (The yaw damper and flight director are not considered parts of the autopilot for purpose of this part). The examiner will select nonprecision approaches that are representative of the type that the applicant is likely to use. The choices must utilize two different types of navigational aids. Some examples of navigational aids for the purpose of this part are: NDB, VOR, LOC, LDA, GPS, or RNAV.

DVD Volume.Segment

Objective. To determine that the applicant:

1. Exhibits adequate knowledge of the elements related to an instrument approach procedure. ....3.2, 3.3, 3.4, 3.7, 3.9, 3.10, 3.13, 3.16, 3.18, 4.10, 6.13, PTS Study Guide Page 2-9
2. Selects and complies with the appropriate instrument approach procedure to be performed. 3.3
3. Establishes two-way communications with ATC, as appropriate, to the phase of flight or approach segment, and uses proper communication phraseology and technique. 2.8, 3.3
4. Selects, tunes, identifies, and confirms the operational status of navigation equipment to be used for the approach procedure. 1.3, 3.3, 3.4, 6.13, Pvt.5.7
5. Complies with all clearances issued by ATC or the examiner. 3.8, 3.10, 7.14
6. Recognizes if any flight instrumentation is inaccurate or inoperative, and takes appropriate action. 1.11, 2.8, 7.11
7. Advises ATC or examiner anytime that the aircraft is unable to comply with a clearance. 7.14
8. Establishes the appropriate aircraft configuration and airspeed considering turbulence and wind shear, and completes the aircraft checklist items appropriate to the phase of the flight. 4.11, 7.2
9. Maintains, prior to beginning the final approach segment, altitude within +/-100 feet, heading within +/-10° and allows less than ¾ scale deflection of the CDI or within +/-10° in the case of an RMI, and maintains airspeed within +/-10 knots. 1.3
10. Applies the necessary adjustments to the published MDA and visibility criteria for the aircraft approach category when required, such as- 3.3, 4.9, 4.10
    a) NOTAMs. 4.10, 7.12
    b) inoperative aircraft and ground navigation equipment. 4.10
    c) inoperative visual aids associated with the landing environment. 4.9, 4.10
    d) NWS reporting factors and criteria. 3.3, 4.10
11. Establishes a rate of descent and track that will ensure arrival at the MDA prior to reaching the MAP with the aircraft continuously in a position from which descent to a landing on the intended runway can be made at a normal rate using normal maneuvers. 1.3, 3.9, 4.10
12. Allows, while on the final approach segment, no more than a three-quarter-scale deflection of the CDI or within 10° in case of an RMI, and maintains airspeed within +/-10 knots of that desired. 1.3, 3.4
13. Maintains the MDA, when reached, within +100 feet, -0 feet to the MAP. 1.3, 3.3
14. Executes the missed approach procedure when the required visual references for the intended runway are not distinctly visible and identifiable at the MAP. 1.3, 3.3, 5.10
15. Executes a normal landing from a straight-in or circling approach when instructed by the examiner.

B. TASK: PRECISION APPROACH (PA)

NOTE: A precision approach, utilizing aircraft NAVAID equipment for centerline and vertical guidance, must be accomplished in simulated or actual instrument conditions to DA/DH.

Objective. To determine that the applicant:

1. Exhibits adequate knowledge of the precision instrument approach procedures. ....3.3, 3.4, 3.5, 3.10, PTS Study Guide Page 2-9
2. Accomplishes the appropriate precision instrument approaches as selected by the examiner. 3.3
3. Establishes two-way communications with ATC using the proper communications phraseology and techniques, as required for the phase of flight or approach segment. 2.8, 3.3
4. Complies, in a timely manner, with all clearances, instructions, and procedures. 3.8, 3.10, 7.14
5. Advises ATC anytime that the applicant is unable to comply with a clearance. 7.14
6. Establishes the appropriate airplane configuration and airspeed/V-speed considering turbulence, wind shear, microburst conditions, or other meteorological and operating conditions. 7.2
7. Completes the aircraft checklist items appropriate to the phase of flight or approach segment, including engine out approach and landing checklists, if appropriate. 4.11
8. Prior to beginning the final approach segment, maintains the desired altitude +/-100 feet, the desired airspeed within +/-10 knots, the desired heading within +/-10°; and accurately tracks radials, courses, and bearings. 1.3
9. Selects, tunes, identifies, and monitors the operational status of ground and airplane navigation equipment used for the approach procedure. 3.3, 3.4, 3.5

DVD Volume.Segment

10. Applies the necessary adjustments to the published DA/DH and visibility criteria for the airplane approach category as required, such as- 3.3, 3.5, 4.9, 4.10
    a) NOTAMs. 4.10, 7.12
    b) inoperative airplane and ground navigation equipment. 3.5, 4.10
    c) inoperative visual aids associated with the landing environment. 3.5, 4.9, 4.10
    d) NWS reporting factors and criteria. 4.10
11. Establishes a predetermined rate of descent at the point where the electronic glide slope begins, which approximates that required for the aircraft to follow the glide slope. 3.5
12. Maintains a stabilized final approach, from the Final Approach Fix to DA/DH allowing no more than three-quarter scale deflection of either the glide slope or localizer indications and maintains the desired airspeed within +/-10 knots. 1.3, 3.4, 3.5
13. A missed approach or transition to a normal landing shall be initiated at Decision Height.
14. Initiates immediately the missed approach when at the DA/DH, and the required visual references for the runway are not unmistakeably visible and identifiable. 1.3, 3.3, 5.10
15. Transitions to a normal landing approach (missed approach for seaplanes) only when the aircraft is in a position from which a descent to a landing on the runway can be made at a normal rate of descent using normal maneuvering.
16. Maintains localizer and glide slope within three-quarter scale deflection of the indicators during the visual descent from DA/DH to a point over the runway where glide slope must be abandoned to accomplish a normal landing.

C. Task: Missed Approach

Objective. To determine that the applicant:

1. Exhibits adequate knowledge of the elements related to missed approach procedures associated with standard instrument approaches. .... 3.3, 3.15, PTS Study Guide Page 2-12
2. Initiates the missed approach promptly by applying power, establishing a climb attitude, and reducing drag in accordance with the aircraft manufacturer's recommendations. .... AFM/POH
3. Reports to ATC beginning the missed approach procedure. 2.6, 2.8
4. Complies with the published or alternate missed approach procedure. 3.3
5. Advises ATC or examiner anytime that the aircraft is unable to comply with a clearance, restriction, or climb gradient. 7.14
6. Follows the recommended checklist items appropriate to the go-around procedure. 4.11
7. Requests, if appropriate, ATC clearance to the alternate airport, clearance limit, or as directed by the examiner. 2.8
8. Maintains the recommended airspeed within +/-10 knots; heading, course, or bearing within +/-10°; and altitude(s) within +/-100 feet during the missed approach procedure. 1.3

D. Task: Circling Approach

Objective. To determine that the applicant:

1. Exhibits adequate knowledge of the elements related to a circling approach procedure. .... 3.15, PTS Study Guide Page 2-11
2. Selects and complies with the appropriate circling approach procedure considering turbulence and wind shear and considering the maneuvering capabilities of the aircraft. 3.15
3. Confirms the direction of traffic and adheres to all restrictions and instructions issued by ATC and the examiner. 3.15
4. Does not exceed the visibility criteria or descend below the appropriate circling altitude until in a position from which a descent to a normal landing can be made. 3.15
5. Maneuvers the aircraft, after reaching the authorized MDA and maintains that altitude within +100 feet, -0 feet and a flight path that permits a normal landing on a runway. The runway selected must be such that it requires at least a 90° change of direction, from the final approach course, to align the aircraft for landing.

E. Task: Landing From A Straight-In Or Circling Approach

Objective. To determine that the applicant:

1. Exhibits adequate knowledge of the elements related to the pilot's responsibilities, and the environmental, operational, and meteorological factors which affect a landing from a straight-in or a circling approach. 3.15
2. Transitions at the DA/DH, MDA, or VDP to a visual flight condition, allowing for safe visual maneuvering and a normal landing. 3.16
3. Adheres to all ATC (or examiner) advisories, such as NOTAMs, wind shear, wake turbulence, runway surface, braking conditions, and other operational considerations.
4. Completes appropriate checklist items for the pre-landing and landing phase. 4.11
5. Maintains positive aircraft control throughout the complete landing maneuver.

VII. AREA OF OPERATION: EMERGENCY OPERATIONS

DVD Volume.Segment

A. Task: Loss Of Communications

Objective. To determine that the applicant exhibits adequate knowledge of the elements related to applicable loss of communication procedures to include:

1. Recognizing loss of communication. 7.9
2. Continuing to destination according to the flight plan....7.8, PTS Study Guide Page 2-66
3. When to deviate from the flight plan. ....7.8, 7.9, PTS Study Guide Page 2-66
4. Timing for beginning an approach at destination. ....7.8, PTS Study Guide Page 2-66

B. Task: One Engine Inoperative During Straight-And-Level Flight And Turns (Multiengine Airplane)

NOTE: This task is only applicable to multiengine airplanes.

C. Task: One Engine Inoperative-Instrument Approach (Multiengine Airplane)

NOTE: This task is only applicable to multiengine airplanes.

D. Task: Approach With Loss Of Primary Flight Instrument Indicators

Note: This approach shall count as one of the required nonprecision approaches.

Objective. To determine that the applicant:

1. Exhibits adequate knowledge of the elements relating to recognizing if primary flight instruments are inaccurate or inoperative, and advise ATC or the examiner....1.11, 7.11, PTS Study Guide Page 2-68
2. Advises ATC or examiner anytime that the aircraft is unable to comply with a clearance....7.14
3. Demonstrates a nonprecision instrument approach without the use of the primary flight instrument using the objectives of the nonprecision approach TASK (AREA OF OPERATION VI, TASK A). 1.11

VIII. AREA OF OPERATION: POSTFLIGHT PROCEDURES

A. TASK: CHECKING INSTRUMENTS AND EQUIPMENT

Objective. To determine that the applicant:

	DVD Volume.Segment
1. Exhibits adequate knowledge of the elements relating to all instrument and navigation equipment for proper operation.	Appendix B
2. Notes all flight equipment for proper operation.	Appendix B
3. Notes all equipment and/or aircraft malfunctions and makes appropriate documentation of improper operation or failure of such equipment.	Appendix B

Supplemental PTS Information

The following information is from the Instrument Rating Practical Test Standards and may useful in your preparation.

General Information

The Flight Standards Service of the Federal Aviation Administration (FAA) has developed this practical test as the standard that shall be used by FAA inspectors and designated pilot examiners when conducting instrument rating—airplane, helicopter, and powered lift practical tests, and instrument proficiency checks for all aircraft. This practical test standard (PTS) shall also be used for the instrument portion of the commercial pilot—airship practical test. Instructors are expected to use this PTS when preparing applicants for practical tests. Applicants should be familiar with this PTS and refer to these standards during their training.

This PTS sets forth the practical test requirements for the addition of an instrument rating to a pilot certificate in airplanes, helicopters, and powered-lift aircraft.

Information considered directive in nature is described in this PTS book in terms, such as "shall" and "must," indicating the actions are mandatory. Guidance information is described in terms, such as "should" and "may," indicating the actions are desirable or permissive, but not mandatory.

The FAA gratefully acknowledges the valuable assistance provided by many industry participants who contributed their time and talent in assisting with the revision of these practical test standards.

This PTS may be purchased from the Superintendent of Documents, U.S. Government Printing Office (GPO), Washington, DC 20402-9325, or from http://bookstore.gpo.gov. This PTS is also available for download, in pdf format, from the Flight Standards Service web site at http://av-info.faa.gov.

This PTS is published by the U.S. Department of Transportation, Federal Aviation Administration, Airman Testing Standards Branch, AFS-630, P.O. Box 25082, Oklahoma City, OK 73125. Comments regarding this handbook should be sent, in e-mail form, to AFS630comments@faa.gov.

Practical Test Standard Concept

Title 14 of the Code of Federal Regulations (14 CFR) part 61 specifies the areas in which knowledge and skill must be demonstrated by the applicant before the issuance of an instrument rating. The CFRs provide the flexibility to permit the FAA to publish practical test standards containing the AREAS OF OPERATION and specific TASKs in which pilot competency shall be demonstrated. The FAA will revise this PTS whenever it is determined that changes are needed in the interest of safety. ***Adherence to the provisions of the regulations and the practical test standards is mandatory for evaluation of instrument pilot applicants.***

Practical Test Book Description

This test book contains the instrument rating practical test standards for airplane, helicopter, and powered lift. It also contains TASK requirements for the addition of airplane, helicopter, or powered lift, if an instrument rating is possessed by the applicant in at least one other aircraft category. Refer to the commercial pilot–airship practical test standard to determine the instrument TASKs required for that practical test. Required TASKs for instrument proficiency checks (PC) are also contained in these practical test standards.

AREAS OF OPERATION are phases of the practical test arranged in a logical sequence within each standard. They begin with preflight preparation and end with postflight procedures. The examiner may conduct the practical test in any sequence that results in a complete and efficient test; ***however, the ground portion of the practical test shall be accomplished before the flight portion.***

TASKs are titles of knowledge areas, flight procedures, or maneuvers appropriate to an AREA OF OPERATION.

The applicant who holds an airplane, helicopter, or powered lift instrument rating will not have to take the entire test when applying for an added rating. The TASKs required for each additional instrument rating are shown in the Rating Task Table.

Applicants for an instrument proficiency check required by 14 CFR section 61.57 must perform to the standards of the TASKs listed in the guidance provided in the Rating Task Table.

NOTE is used to emphasize special considerations required in the AREA OF OPERATION or TASK.

The Objective lists the important elements that must be satisfactorily performed to demonstrate competency in a TASK. The Objective includes:

1) specifically what the applicant should be able to do;
2) the conditions under which the TASK is to be performed; and
3) the acceptable standards of performance.

Abbreviations

14 CFR	Title 14 of the Code of Federal Regulations
ADF	Automatic Direction Finder
ADM	Aeronautical Decision Making
AIRMETS	Airman's Meteorological Information
APV	Approach with Vertical Guidance
ATC	Air Traffic Control
ATIS	Automatic Terminal Information Service
ATS	Air Traffic Service
CDI	Course Deviation Indicator
CFIT	Controlled Flight into Terrain
CRM	Cockpit Resource Management
DA/DH	Decision Altitude/Decision Height
DH	Decision Height
DME	Distance Measuring Equipment
DP	Departure Procedures
FAA	Federal Aviation Administration
FDC	Flight Data Center
FMS	Flight Management System
FSDO	Flight Standards District Office
GLS	GNSS Landing System
GNSS	Global Navigation Satellite System
GPO	Government Printing Office
GPS	Global Positioning System
GPWS	Ground Proximity Warning System
IAP	Instrument Approach Procedures
IFR	Instrument Flight Rules
ILS	Instrument Landing System
IMC	Instrument Meteorological Conditions
LAHSO	Land and Hold Short Operations
LCD	Liquid Crystal Display
LDA	Localizer-type Directional Aid
LED	Light Emitting Diode
LOC	ILS Localizer
LORAN	Long Range Navigation
MAP	Missed Approach Point
MDA	Minimum Descent Attitude
MLS	Microwave Landing System
NAVAID	Navigation Aid
NDB	Non-directional Beacon (Automatic Direction Finder)
NOTAM	Notice to Airmen
NPA	Nonprecision Approach
NWS	National Weather Service
PA	Precision Approach
PC	Proficiency Check
PTS	Practical Test Standard
RAIM	Receiver Autonomous Integrity Monitoring
RMI	Radio Magnetic Indicator
RNAV	Area Navigation
SAS	Stability Augmentation System
SDF	Simplified Directional Facility
SIGMETS	Significant Meteorological Advisory
STAR	Standard Terminal Arrival
TCAS	Traffic Alert and Collision Avoidance System
VDP	Visual Descent Point
VHF	Very High Frequency
VNAV	Vertical Navigation
VOR	Very High Frequency Ominidirectional Range

Use of the Practical Test Standards

The instrument rating practical test standards are designed to evaluate competency in both knowledge and skill.

The FAA requires that all practical tests be conducted in accordance with the appropriate practical test standards and the policies set forth in the INTRODUCTION. Instrument rating applicants shall be evaluated in **ALL** TASKs included in the AREAS OF OPERATION of the appropriate practical test standard (unless noted otherwise).

In preparation for each practical test, the examiner shall develop a written "plan of action" for each practical test. The "plan of action" is a tool, for the sole use of the examiner, to be used in evaluating the applicant. The plan of action need not be grammatically correct or in any formal format. The plan of action must contain all of the required AREAS OF OPERATION and TASKs and any optional TASKs selected by the examiner. The "plan of action" shall incorporate one or more scenarios that will be used during the practical test. The examiner should try to include as many of the TASKs into the scenario portion of the test as possible, but maintain the flexibility to change due to unexpected situations as they arise and still result in an efficient and valid test. ***Any TASK selected for evaluation during a practical test shall be evaluated in its entirety.***

The examiner is not required to follow the precise order in which the AREAS OF OPERATION and TASKs appear in this book. The examiner may change the sequence or combine TASKs with similar Objectives to have an orderly and efficient flow of the practical test. For example, holding procedures may be combined with an approach or missed approach procedures if a holding entry is part of the procedure.

The TASKs apply to airplanes, helicopters, powered lift, and airships. In certain instances, NOTEs describe differences in the performance of a TASK by an "airplane'' applicant, "helicopter" applicant, or "powered lift" applicant. When using the practical test standards, the examiner must evaluate the applicant's knowledge and skill in sufficient depth to determine that the standards of performance listed for all TASKs are met.

All TASKs in these practical test standards are required for the issuance of an instrument rating in airplanes, helicopters, and powered lift. However, when a particular element is not appropriate to the aircraft, its equipment, or operational capability, that element may be omitted. Examples of these element exceptions would be high altitude weather phenomena for helicopters, integrated flight systems for aircraft not so equipped, or other situations where the aircraft or operation is not compatible with the requirement of the element.

Special Emphasis Areas

Examiners shall place special emphasis upon areas of aircraft operations considered critical to flight safety. Among these are:

1) positive aircraft control;
2) positive exchange of the flight controls procedure (who is flying the aircraft);
3) stall/spin awareness;
4) collision avoidance;
5) wake turbulence avoidance;
6) LAHSO;
7) runway incursion avoidance;
8) CFIT;
9) ADM and risk management;
10) checklist usage; and
11) other areas deemed appropriate to any phase of the practical test.

Although these areas may not be specifically addressed under each TASK, they are essential to flight safety and will be evaluated during the practical test.

Practical Test Prerequisites: Instrument Rating

An applicant for an instrument rating practical test is required by 14 CFR part 61 to:

1) hold at least a current private pilot certificate with an aircraft rating appropriate to the instrument rating sought;
2) pass the appropriate instrument rating knowledge test since the beginning of the 24th month before the month in which the practical test is taken;
3) obtain the applicable training and aeronautical experience prescribed for the instrument rating sought;
4) hold at least a current third-class medical certificate;
5) be able to read, speak, write, and understand the English language; and
6) obtain a written statement from an authorized flight instructor certifying that the applicant has been given flight training in preparation for the practical test within 60 days preceding the date of application. The statement shall also state that the instructor finds the applicant competent to pass the practical test and that the applicant has satisfactory knowledge of the subject area(s) in which a deficiency was indicated by the Airman Knowledge Test Report.

If there are questions concerning English language requirements, refer to AC 60-28, English Language Skill Standards Required by 14 CFR parts 61, 63, and 65, or your local FSDO. English language requirements should be determined to be met prior to beginning the practical test.

Aircraft and Equipment Required for the Practical Test

The instrument rating applicant is required by 14 CFR part 61 to provide an airworthy, certificated aircraft for use during the practical test. Its operating limitations must not prohibit the TASKs required on the practical test. Flight instruments are those required for controlling the aircraft without outside references. The required radio equipment is that which is necessary for communications with ATC, and for the performance of two of the following nonprecision approaches: VOR, NDB, GPS, LOC, LDA, SDF, or RNAV and one precision approach: ILS, GLS, or MLS. GPS equipment must be instrument certified and contain the current database. Note: APV approaches may be substituted only for nonprecision approaches in this standard. An APV approach shall not be used in lieu of the required precision approach.

Modern technology has introduced into aviation a new method of displaying flight instruments, such as Electronic Flight Instrument Systems, Integrated Flight Deck displays, and others. For the purpose of the practical test standards, any flight instrument display that utilizes LCD or picture tube like displays will be referred to as "Electronic Flight Instrument Display." Aircraft equipped with this technology may or may not have separate backup flight instruments installed. The abnormal or emergency procedure for loss of the electronic flight instrument display appropriate to the aircraft will be evaluated in the Loss of Primary Instruments TASK. The loss of the primary electronic flight instrument display must be tailored to failures that would normally be encountered in the aircraft. If the aircraft is capable, total failure of the electronic flight instrument display, or a supporting component, with access only to the standby flight instruments or backup display shall be evaluated.

The applicant is required to provide an appropriate view limiting device that is acceptable to the examiner. This device shall be used during all testing that requires testing "solely by reference to instruments." This device must prevent the applicant from having visual reference outside the aircraft, but not prevent the examiner from having visual reference outside the aircraft. A procedure should be established between the applicant and the examiner as to when and how this device should be donned and removed and this procedure briefed before the flight.

The applicant is expected to utilize an autopilot and/or flight management system (FMS), if properly installed, during the instrument practical test to assist in the management of the aircraft. The examiner is expected to test the applicant's knowledge of the systems that are installed and operative during the oral and flight portions of the practical test. The applicant will be required to demonstrate the use of the autopilot and/or FMS during one of the nonprecision approaches.

If the practical test is conducted in the aircraft, and the aircraft has an operable and properly installed GPS, the applicant must demonstrate GPS approach proficiency when asked. If the applicant has contracted for training in an approved course that includes GPS training in the system that is installed in the airplane/simulator/FTD and the airplane/simulator/FTD used for the checking/testing has the same system properly installed and operable, the applicant must demonstrate GPS approach proficiency. When a practical test is conducted for a 14 CFR part 135 operator, the operator's approved training program is the controlling authority.

To obtain an **instrument rating with multiengine privileges**, an applicant must demonstrate competency in a multiengine airplane not limited to center thrust. The multiengine airplane that is used to obtain unlimited multiengine privileges must have a published V_{MC} speed established by the manufacturer, and produce an asymmetrical thrust configuration with the loss of one or more engines. If an instrument flight test is conducted in a multiengine airplane limited to center thrust, a limitation shall be placed on the applicant's certificate: INSTRUMENT RATING, AIRPLANE MULTIENGINE, LIMITED TO CENTER THRUST.

Use of FAA-Approved Flight Simulator or Flight Training Device

An airman applicant for instrument rating certification is authorized to use an FAA-qualified and approved flight simulator or flight training device, to complete certain flight TASK requirements listed in this practical test standard.

When flight TASKs are accomplished in an aircraft, certain TASK elements may be accomplished through "simulated" actions in the interest of safety and practicality, but when accomplished in a flight simulator or flight training device, these same actions would not be "simulated." For example, when in an aircraft, a simulated engine fire may be addressed by retarding the throttle to idle, simulating the shutdown of the engine, simulating the discharge of the fire suppression agent, if applicable, simulating the disconnect of associated electrical, hydraulic, and pneumatics systems. However, when the same emergency condition is addressed in a flight simulator or flight training device, all TASK elements must be accomplished as would be expected under actual circumstances.

Similarly, safety of flight precautions taken in the aircraft for the accomplishment of a specific maneuver or procedure (such as limiting altitude in an approach to stall or setting maximum airspeed for an engine failure expected to result in a rejected takeoff) need not be taken when a flight simulator or flight training device is used.

It is important to understand that whether accomplished in an aircraft, flight simulator or flight training device, all TASKs and elements for each maneuver or procedure shall have the same performance standards applied equally for determination of overall satisfactory performance.

The applicant must demonstrate all of the instrument approach procedures required by 14 CFR part 61. At least one instrument approach procedure must be demonstrated in an airplane, helicopter, or powered lift as appropriate. One precision and one nonprecision approach not selected for actual flight demonstration may be performed in flight simulators or flight training devices that meet the requirements of appendix 1 of this practical test standard.

Flight Instructor Responsibility

An appropriately rated flight instructor is responsible for training the instrument rating pilot applicant to acceptable standards in all subject matter areas, procedures, and maneuvers included in the TASKs within the appropriate instrument rating practical test standard.

Because of the impact of their teaching activities in developing safe, proficient pilots, flight instructors should exhibit a high level of knowledge, skill, and the ability to impart that knowledge and skill to students. Additionally, the flight instructor must certify that the applicant is able to perform safely as an instrument pilot and is competent to pass the required practical test.

Throughout the applicant's training, the flight instructor is responsible for emphasizing the performance of effective visual scanning, collision avoidance, and runway incursion avoidance procedures. These areas are covered, in part, in AC 90-48, Pilot's Role in Collision Avoidance; FAA-H-8083-3, Airplane Flying Handbook; FAA-H-8083-25, Pilot's Handbook of Aeronautical Knowledge; and the Aeronautical Information Manual.

Examiner[1] Responsibility

The examiner conducting the practical test is responsible for determining that the applicant meets the acceptable standards of knowledge and skill of each TASK within the appropriate practical test standard. Since there is no formal division between the "oral" and "skill" portions of the practical test, this becomes an ongoing process throughout the test. To avoid unnecessary distractions, oral questioning should be used judiciously at all times, especially during the flight portion of the practical test.

Examiners shall test to the greatest extent practicable the applicant's correlative abilities rather than mere rote enumeration of facts throughout the practical test.

If the examiner determines that a TASK is incomplete, or the outcome uncertain, the examiner may require the applicant to repeat that TASK, or portions of that TASK. This provision has been made in the interest of fairness and does not mean that instruction, practice, or the repeating of an unsatisfactory TASK is permitted during the certification process.

During the flight portion of the practical test, the examiner shall evaluate the applicant's use of visual scanning, and collision avoidance procedures, when appropriate. Except for takeoff and landing, all TASKs shall be conducted solely by reference to instruments under actual or simulated instrument flight conditions.

The examiner may not assist the applicant in the management of the aircraft, radio communications, navigational equipment, and navigational charts. In the event the test is conducted in an aircraft operation requiring a crew of two, the examiner may assume the duties of the second in command. Helicopters certified for IFR operations must be flown using two pilots or a single pilot with an approved autopilot or SAS. Therefore, when conducting practical tests in a helicopter (without autopilot, SAS, or copilot), examiners may act as an autopilot (e.g., hold heading and altitude), when requested, to allow applicants to tune radios, select charts, etc. Examiners may perform the same functions as an autopilot but should not act as a copilot performing more extensive duties. The examiner shall remain alert for other traffic at all times. The examiner shall use proper ATC terminology when simulating ATC clearances.

[1]The word "examiner" is used throughout the standard to denote either the FAA inspector or FAA designated pilot examiner who conducts an official practical test.

Satisfactory Performance

Satisfactory performance to meet the requirements for certification is based on the applicant's ability to safely:

1) perform the TASKs specified in the AREAS OF OPERATION for the certificate or rating sought within the approved standards;
2) demonstrate mastery of the aircraft with the successful outcome of each TASK performed never seriously in doubt;

3) demonstrate satisfactory proficiency and competency within the approved standards;

4) demonstrate sound judgment and ADM; and

5) demonstrate single-pilot competence if the aircraft is type certificated for single-pilot operations.

Unsatisfactory Performance

The tolerances represent the performance expected in good flying conditions. If, in the judgment of the examiner, the applicant does not meet the standards of performance of any TASK performed, the associated AREA OF OPERATION is failed and therefore, the practical test is failed.

Note: The tolerances stated in this standard are intended to be used as a measurement of the applicant's ability to operate in the instrument environment. They provide guidance for examiners to use in judging the applicant's qualifications. The regulations governing the tolerances for operation under Instrument Flight Rules are established in 14 CFR part 91.

The examiner or applicant may discontinue the test at any time when the failure of an AREA OF OPERATION makes the applicant ineligible for the certificate or rating sought. ***The test may be continued ONLY with the consent of the applicant.*** If the test is discontinued, the applicant is entitled credit for only those AREAS OF OPERATION and their associated TASKs satisfactorily performed. However, during the retest, and at the discretion of the examiner, any TASK may be reevaluated, including those previously passed.

Typical areas of unsatisfactory performance and grounds for disqualification are:

1) Any action or lack of action by the applicant that requires corrective intervention by the examiner to maintain safe flight.

2) Failure to use proper and effective visual scanning techniques, when applicable, to clear the area before and while performing maneuvers.

3) Consistently exceeding tolerances stated in the Objectives.

4) Failure to take prompt corrective action when tolerances are exceeded.

When a notice of disapproval is issued, the examiner shall record the applicant's unsatisfactory performance in terms of the AREA OF OPERATION and specific TASK(s) not meeting the standard appropriate to the practical test conducted. The AREA(s) OF OPERATION/TASK(s) not tested and the number of practical test failures shall also be recorded. If the applicant fails the practical test because of a special emphasis area, the Notice of Disapproval shall indicate the associated TASK. For example, AREA OF OPERATION VI, TASK E, Landing From a Straight-in or Circling Approach, failure to avoid runway incursion.

Letter of Discontinuance

When a practical test is discontinued for reasons other than unsatisfactory performance (i.e., equipment failure, weather, or illness) FAA Form 8700-1, Airman Certificate and/or Rating Application, and, if applicable, the Airman Knowledge Test Report, shall be returned to the applicant. The examiner at that time shall prepare, sign, and issue a Letter of Discontinuance to the applicant. The Letter of Discontinuance should identify the AREAS OF OPERATION and their associated TASKs of the practical test that were successfully completed. The applicant shall be advised that the Letter of Discontinuance shall be presented to the examiner when the practical test is resumed, and made part of the certification file.

Aeronautical Decision Making And Risk Management

The examiner shall evaluate the applicant's ability throughout the practical test to use good aeronautical decision making procedures in order to evaluate risks. The examiner shall accomplish this requirement by developing scenarios that incorporate as many TASKs as possible to evaluate the applicants risk management in making safe aeronautical decisions. For example, the examiner may develop a scenario that incorporates weather decisions and performance planning.

The applicant's ability to utilize all the assets available in making a risk analysis to determine the safest course of action is essential for satisfactory performance. The scenarios should be realistic and within the capabilities of the aircraft used for the practical test.

Single-Pilot Resource Management

Single-Pilot Resource Management refers to the effective use of ALL available resources: human resources, hardware, and information. It is similar to Crew Resource Management (CRM) procedures that are being emphasized in multi-crewmember operations except that only one crewmember (the pilot) is involved. Human resources "...includes all other groups routinely working with the pilot who are involved in decisions that are required to operate a flight safely. These groups include, but are not limited to: dispatchers, weather briefers, maintenance personnel, and air traffic controllers." Pilot Resource Management is not a single TASK; it is a set of skill competencies that must be evident in all TASKs in this practical test standard as applied to single-pilot operation.

Applicant's Use of Checklists

Throughout the practical test, the applicant is evaluated on the use of an appropriate checklist. Proper use is dependent on the specific TASK being evaluated. The situation may be such that the use of the checklist, while accomplishing elements of an Objective, would be either unsafe or impractible, especially in a single-pilot operation. In this case, a review of the checklist after the elements have been accomplished would be appropriate. Division of attention and proper visual scanning should be considered when using a checklist.

Use of Distractions During Practical Tests

Numerous studies indicate that many accidents have occurred when the pilot has been distracted during critical phases of flight. To evaluate the pilot's ability to utilize proper control technique while dividing attention both inside and/or outside the cockpit, the examiner shall cause a realistic distraction during the flight portion of the practical test to evaluate the applicant's ability to divide attention while maintaining safe flight.

Positive Exchange of Flight Controls

During flight, there must always be a clear understanding between pilots of who has control of the aircraft. Prior to flight, a briefing should be conducted that includes the procedure for the exchange of flight controls. A positive three-step process in the exchange of flight controls between pilots is a proven procedure and one that is strongly recommended.

When one pilot wishes to give the other pilot control of the aircraft, he or she will say, "You have the flight controls." The other pilot acknowledges immediately by saying, "I have the flight controls." The first pilot again says, "You have the flight controls." When control is returned to the first pilot, follow the same procedure. A visual check is recommended to verify that the exchange has occurred. There should never be any doubt as to who is flying the aircraft.

Emphasis on Attitude Instrument Flying and Emergency Instrument Procedures

The FAA is concerned about numerous fatal aircraft accidents involving spatial disorientation of instrument-rated pilots who have attempted to control and maneuver their aircraft in clouds with inoperative primary flight instruments (gyroscopic heading and/or attitude indicators) or loss of the primary electronic flight instruments display.

AREA OF OPERATION IV requires the evaluation of basic instrument flight maneuvers under both full-panel and references to backup primary flight instruments/electronic flight instrument displays. These maneuvers are described in detail in FAA-H-8083-15, Instrument Flying Handbook. Examiners should determine that the applicant demonstrates competency in either the PRIMARY and SUPPORTING or the CONTROL and PERFORMACE CONCEPT method of instrument flying. Either attitude instrument flying method is described in FAA-H-8083-15 and is recommended by the FAA because it requires specific knowledge and interpretation of each individual instrument during training.

The FAA has stressed that it is imperative for instrument pilots to acquire and maintain adequate instrument skills and that they be capable of performing instrument flight with the use of the backup systems installed in the aircraft. Many light aircraft operated in IMC are not equipped with dual, independent, gyroscopic heading and/or attitude indicators and in many cases are equipped with only a singlevacuum source. Technically advanced aircraft may be equipped with backup flight instruments or an additional electronic flight display that is not located directly in front of the pilot.

The instrument rating practical test standards place emphasis on and requires the demonstrations of a nonprecision instrument approach without the use of the primary flight instruments or electronic flight instrument display. A nonprecision approach without the use of the primary flight instruments/electronic flight instrument display is considered one of the most demanding situations that could be encountered. If applicants can master this situation, they can successfully complete a less difficult precision approach. If an actual approach in IMC becomes necessary without the aid of the primary flight instruments/ electronic flight instrument display, a less difficult precision approach should be requested, if available. Sound judgment would normally dictate such requests. However, the instrument practical test requires that a nonprecision approach be performed without the use of the primary flight instruments/electronic flight instrument display.

Applicants may have an unfair advantage during performance of the TASK using the backup flight instruments during an instrument approach due to the location of the magnetic compass in some aircraft. When crosschecking the magnetic compass heading, a view of the runway or other visual clue may be sighted. It is the examiner's responsibility to determine if the applicant is receiving visual clues from outside the cockpit. If an examiner feels that the applicant is receiving visual clues, the examiner may devise other options to limit the applicant's view. By no means shall the examiner limit his or her view as the safety pilot.

Rating Task Table

ADDITIONAL INSTRUMENT RATING DESIRED				
AREA OF OPERATION	Required TASKS are indicated by either the TASK letter(s) that apply(s) or an indication that all or none of the TASKS must be tested.			
	IA	IH	IPL	IPC
I	NONE	NONE	NONE	NONE
II	A, C	A, C	A, C	NONE
III	NONE	NONE	NONE	C
IV	ALL	ALL	ALL	B
V	NONE	NONE	NONE	ALL
VI	ALL	ALL	ALL	ALL*
VII	ALL**	ALL**	ALL**	B, C, D
VIII	ALL	ALL	ALL	ALL

Legend

IA - Instrument airplane

IH - Instrument helicopter

IPL - Instrument powered lift

IPC - Instrument Proficiency check

NOTE: Except as noted, all TASKs are required for *initial issuance* of an instrument rating.

* TASK D, Circling Approach, is applicable *only to* the *airplane* category.

** TASKS B and C are applicable *only* to *multiengine airplanes.*

Instrument Proficiency Check. 14 CFR part 61.57(d) sets forth the requirements for an instrument proficiency check. The person giving that check shall use the standards and procedures contained in this PTS when administering the check. A representative number of TASKs, as determined by the examiner/instructor, must be selected to assure the competence of the applicant to operate in the IFR environment. As a minimum, the applicant must demonstrate the ability to perform the TASKs as listed in the above chart. The person giving the check should develop scenarios to assess the pilot's ADM and risk management skills during the IPC.

Applicant's Practical Test Checklist

APPOINTMENT WITH EXAMINER:

EXAMINER'S NAME____________________________

LOCATION __________________________________

DATE/TIME __________________________________

ACCEPTABLE AIRCRAFT

- ❑ View-limiting device
- ❑ Aircraft Documents: Airworthiness Certificate
- ❑ Registration Certificate
- ❑ Rating Limitations
- ❑ Aircraft Maintenance Records: Airworthiness Inspections

PERSONAL EQUIPMENT

- ❑ Current Aeronautical Charts
- ❑ Computer and Plotter
- ❑ Flight Plan Form
- ❑ Flight Logs
- ❑ Current AIM

PERSONAL RECORDS

- ❑ Identification - Photo/Signature ID
- ❑ Pilot Certificate
- ❑ Medical Certificate
- ❑ Completed FAA Form 8710-1, Application for an Airman Certificate and/or Rating
- ❑ Airman Knowledge Test Report
- ❑ Logbook with Instructor's Endorsement
- ❑ Notice of Disapproval (if applicable)
- ❑ Approved School Graduation Certificate (if applicable)
- ❑ Examiner's Fee (if applicable)

Examiner's Practical Test Checklist

APPLICANT'S NAME____________________________

LOCATION____________________________________

DATE/TIME____________________________________

I. PREFLIGHT PREPARATION

- ❑ A. Weather Information
- ❑ B. Cross-Country Flight Planning

II. PREFLIGHT PROCEDURES

- ❑ A. Aircraft Systems Related to IFR Operations
- ❑ B. Aircraft Flight Instruments and Navigation Equipment
- ❑ C. Instrument Cockpit Check

III. AIR TRAFFIC CONTROL CLEARANCES AND PROCEDURES

- ❑ A. Air Traffic Control Clearances
- ❑ B. Compliance with Departure, En Route, and Arrival Procedures and Clearances
- ❑ C. Holding Procedures

IV. FLIGHT BY REFERENCE TO INSTRUMENTS

- ❑ A. Basic Instrument Flight Maneuvers
- ❑ B. Recovery from Unusual Flight Attitudes

V. NAVIGATION SYSTEMS

- ❑ A. Intercepting and Tracking Navigational Systems and DME Arcs

VI. INSTRUMENT APPROACH PROCEDURES

- ❑ A. Nonprecision Approach (NPA)
- ❑ B. Precision Approach (PA)
- ❑ C. Missed Approach
- ❑ D. Circling Approach
- ❑ E. Landing from a Straight-in or Circling Approach

VII. EMERGENCY OPERATIONS

- ❑ A. Loss of Communications
- ❑ B. One Engine Inoperative During Straight-and-Level Flight and Turns (Multiengine Airplane)
- ❑ C. One Engine Inoperative—Instrument Approach (Multiengine Airplane)
- ❑ D. Loss of Primary Flight Instrument Indicators

VIII. POSTFLIGHT PROCEDURES

- ❑ A. Checking Instruments and Equipment

PTS Appendix 1 - Task vs. Simulation Device Credit

Examiners conducting the instrument rating practical tests with flight simulation devices should consult appropriate documentation to ensure that the device has been approved for training, testing, or checking. The documentation for each device should reflect that the following activities have occurred:

1) The device must be evaluated, determined to meet the appropriate standards, and assigned the appropriate qualification level by the National Simulator Program Manager. The device must continue to meet qualification standards through continuing evaluations as outlined in the appropriate advisory circular (AC). For airplane flight training devices (FTDs), AC 120-45 (as amended), Airplane Flight Training Device Qualifications, will be used. For simulators, AC 120-40 (as amended), Airplane Simulator Qualification, will be used.

2) The FAA must approve the device for training, testing, and checking the specific flight TASKs listed in this appendix.

3) The device must continue to support the level of student or applicant performance required by this practical test standard.

NOTE: Users of the following chart are cautioned that use of the chart alone is incomplete. The description and Objective of each TASK as listed in the body of the practical test standard, including all NOTEs, must also be incorporated for accurate simulation device use.

Use of Chart

X Creditable.

A Creditable if appropriate systems are installed and operating.

NOTES:

- Level 1 FTDs that have been issued a letter authorizing their use by the FAA Administrator, may continue to be used only for those TASKs originally found acceptable. Use of Level 1, 2, or 3 FTDs may not be used for aircraft requiring a type rating.
- If a FTD or a simulator is used for the practical test, the instrument approach procedures conducted in that FTD or simulator are limited to one precision and one nonprecision approach procedure.
- Postflight procedures means, closing flight plans, checking for discrepancies and malfunctions, and noting them on a log or maintenance form.

FLIGHT TASK	FLIGHT SIMULATION DEVICE LEVEL											
Areas of Operation	**1**	**2**	**3**	**4**	**5**	**6**	**7**		**A**	**B**	**C**	**D**
II. Preflight Procedures												
C. Instrument Cockpit Check *		A	X	A	A	X	X		X	X	X	X
III. Air Traffic Control Clearances and Procedures												
A. Air Traffic Control Clearances *		A	X	A	A	X	X		X	X	X	X
B. Departure, En Route and Arrival Clearances *			X			X	X		X	X	X	X
C. Holding Procedures			X			X	X		X	X	X	X
IV. Flight by Reference to Instruments												
A. Basic Instrument Flight Maneuvers			X			X	X		X	X	X	X
B. Recovery from Unusual Flight Attitudes							X		X	X	X	X
V. Navigation Systems												
A. Intercepting and Tracking Navigational Systems and DME ARCS		A	X		A	X	X		X	X	X	X
VI. Instrument Approach Procedures												
A. Nonprecision Approach (NPA)			X			X	X		X	X	X	X
B. Precision Approach (PA)			X			X	X		X	X	X	X
C. Missed Approach			X			X	X		X	X	X	X
D. Circling Approach									X	X	X	X
E. Landing from a Straight-in or Circling Approach										X	X	X
VII. Emergency Operations **												
A. Loss of Communications		X	X			X	X		X	X	X	X
B. One Engine Inoperative during Straight-and-Level Flight and Turns (Multiengine Airplane)			X			X	X		X	X	X	X
C. One Engine Inoperative—Instrument Approach (Multiengine Airplane)									X	X	X	X
D. Loss of Gyro Attitude and/or Heading Indicators						X	X		X	X	X	X
VIII. Postflight Procedures												
A. Checking Instruments and Equipment		A	X		A	X	X		X	X	X	X

* Aircraft required for those items that cannot be checked using a flight training device or flight simulator.

** Multiengine - Single Engine

(Intentionally Left Blank)

Section 2 - Instrument Rating DVD Study Guide

The following pages should be used as reinforcing material while reviewing the various DVD volumes.

Please remember these notes cannot serve as a substitute for the instruction contained in the video. They are intended to reinforce essential material from the *What You Should Know* DVD Series and will assist you in learning these subjects.

Volume 1 - Instrument Flying Fundamentals

Requirements

1) The requirements for the instrument rating are listed in 14 CFR §61.65. This regulation includes a description of both flight and ground instruction requirements. In order to be eligible for the instrument rating, you must:
   a) Hold a current private pilot certificate and have at least 50 hours of cross-country flight time as pilot in command. 40 hours of actual or simulated instrument flight time is necessary. Of these 40 hours, 15 must be must be given by an authorized instrument instructor in the aircraft category for which the instrument rating is sought. 20 hours may be in an approved flight simulator or flight training device. Of these 20 hours, up to 10 may be in an approved PC-based aviation training device. One instrument cross-country of at least 250 nautical miles with three different kinds of instrument approaches at three different airports must be completed.
   b) Pass the knowledge test. This examination covers **Regulations, Aircraft Instruments, Radio Navigation, Flight Planning, En Route and Approach Procedures,** and **Weather**. Practice knowledge test questions are available on each DVD. Additional practice may be obtained through **sportys.com/faatest**.
   c) Pass the practical and oral portions of the examination. The criteria for these tests are discussed in the **Instrument Rating Practical Test Standards** found in Section 1 of this book.

The Practical Test

1) The practical examination is structured to resemble an actual flight under instrument conditions. You must have a thorough knowledge of aviation weather, the airplane systems and instruments, ATC system, and cross-country planning, including a knowledge of the aircraft's performance characteristics.
   a) Refer to Section 1 of this study guide for additional details regarding the practical test.

Instruments

PITCH INSTRUMENTS	BANK INSTRUMENTS	POWER INSTRUMENTS
Airspeed Indicator	Heading Indicator	Airspeed Indicator
Attitude Indicator	Attitude Indicator	Tachometer
Altimeter	Turn indicator	Manifold Pressure Gauge
Vertical Speed Indicator		

1) Pitch
   a) The attitude indicator gives a direct indication of pitch attitude.
   b) Given a constant power setting, the altimeter, airspeed indicator, and vertical speed indicator indirectly indicate pitch attitude.
   c) Pitch corrections should be made using the attitude indicator, and should be made as half (100 ft. or less correction), full (more than 100 ft. correction), and one and one half bar width corrections. Pitch corrections to maintain level flight should be smaller than during VFR flying.
   d) The Vertical Speed Indicator is used only as a secondary indication due to the lag time associated with its readings, and is not required for not-for-hire IFR flight.

2) Bank
   a) The attitude indicator gives direct indication of bank attitude.
   b) The heading indicator and turn indicator indirectly indicate bank attitude.
   c) The heading indicator is the primary bank indication instrument in straight and level flight. The magnetic compass should be used in case of heading indicator failure.
   d) The turn coordinator measures the rate of turn and rate of roll, which is movement around the longitudinal axis. It does not give a direct indication of the bank angle.
   e) Small bank deviations are most easily detected by the banking scale and center index on the attitude indicator.
   f) For heading corrections of 5° or less, use rudder pressure to return to course. For corrections of more than 5°, a coordinated turn should be made.

Fundamental Skills

1) **Instrument Cross-Check** (Instrument Scan).
   a) Scanning Errors:
      i) **Omission**, or the neglect of an instrument or instruments, usually occurs when an attitude change is made without checking performance instruments. For example, leveling off from a climb by establishing level flight on the attitude indicator, and ignoring the altimeter.
      ii) **Emphasis** of one instrument over other instruments is also detrimental to precision flight.
      iii) **Fixation** on a particular instrument tends to occur when approaching an altitude during climbs (altimeter) or approaching a heading during turns (heading indicator).
2) **Instrument Interpretation**.
3) **Aircraft Control**.

Illusions in Flight

1) Illusions in flight can lead to spatial disorientation and/or landing errors. Illusions rank among the most common factors cited as contributing to fatal aircraft accidents.
2) **Illusions Leading to Spatial Disorientation** can be prevented only by visual reference to reliable flight instruments.
   a) **The leans** - An abrupt correction of a banked attitude, which has been entered too slowly to stimulate the motion sensing system in the inner ear, creates the illusion of banking in the opposite direction.
   b) **Coriolis illusion** - An abrupt head movement in a prolonged constant rate turn that has ceased stimulating the motion sensing system can create the illusion of rotation or movement in an entirely different axis.
   c) **Graveyard spin** - A proper recovery from a spin that has ceased stimulating the motion sensing system can create the illusion of spinning in the opposite direction.
   d) **Graveyard spiral** - An observed loss of altitude during a coordinated constant-rate turn that has ceased stimulating the motion sensing system can create the illusion of being in a descent with the wings level.
   e) **Somatogravic illusion** - A rapid acceleration during takeoff can create the illusion of being in a nose up attitude.
   f) **Inversion illusion** - An abrupt change from climb to straight and level flight can create the illusion of tumbling backwards.
   g) **Elevator illusion** - An abrupt upward vertical acceleration, usually by an updraft, can create the illusion of being in a climb.
   h) **False horizon** - Sloping cloud formations, an obscured horizon, a dark scene spread with ground lights and stars, and certain geometric patterns of ground lights can create illusions of not being aligned correctly with the actual horizon.
   i) **Autokinesis** - In the dark, a static light will appear to move about when stared at for many seconds.

3) **Illusions Leading to Landing Errors** can be prevented by anticipating them during approaches, using electronic glide slope or VASI systems when available, and maintaining proficiency in landing procedures.

   a) **Runway width illusion** -- A narrower-than-usual runway can create the illusion that the aircraft is at a higher altitude than it actually is.

   b) **Runway and terrain slopes illusion** -- An upsloping runway, upsloping terrain, or both can create the illusion that the aircraft is at a higher altitude than it actually is.

   c) **Featureless terrain illusion** -- An absence of ground features, as when landing over water, darkened areas, and terrain made featureless by snow, can create the illusion that the aircraft is at a higher altitude than it actually is.

   d) **Atmospheric illusion** -- Rain on the windscreen can create the illusion of greater height, and atmospheric haze can create the illusion of being at a greater distance from the runway.

   e) **Ground lighting illusions** -- Lights along a straight path, such as a road and lights on moving trains can be mistaken for runway and approach lights. Bright runway and approach lighting systems, especially where few lights illuminate the surrounding terrain, may create the illusion of less distance to the runway.

Volume 2 - Air Traffic Control and IFR

Separation

1) Vertical Separation
   a) Up to but not including FL 290, vertical separation must be at least 1,000 ft.
   b) At and above FL 290, vertical separation is increased to 2,000 ft. unless the airspace is designated for Reduced Vertical Separation Minimum (RSVM).
      i) The airspace from FL290 to FL410 over the United States is designated as RVSM airspace.
      ii) RSVM requires special equipment and training and allows 1,000 ft. vertical separation.
2) Horizontal Separation
   a) When horizontal separation is determined by radar, the actual separation distances depend on the type of radar and the distance of the aircraft from the radar antenna.
   b) For IFR flights made in areas outside of radar control, separation is based on position reports. This method is also used to control aircraft in the event of a radar failure. When reporting, the following information is necessary:
      i) Airplane Identification
      ii) Name of reporting point
      iii) Time over the reporting point
      iv) Altitude or flight level
      v) Type of flight plan -- (Omitted when reporting to TRACON or ARTCC.)
      vi) Name of and estimated time over **next** compulsory reporting point
      vii) Name only of the **next succeeding** compulsory reporting point
      viii) Pertinent Remarks
3) Wake turbulence from a heavy airplane (an airplane capable of more than a 255,000-lb.takeoff weight or as designated by the administrator) may require a separation distance of as much as six miles.
4) Collision avoidance is always the pilot's responsibility. 14 CFR §91.113(a) states: "When weather conditions permit, regardless of whether an operation is conducted under VFR or IFR, vigilance shall be maintained by each person operating an aircraft so as to see and avoid other aircraft."
5) Separation during departure from a nontower airport may depend on the pilot adhering to **release time** and **clearance void time** restrictions given by ATC as part of the clearance.
   a) A release time is a departure restriction issued by ATC, specifying the earliest time an aircraft may depart.
   b) A pilot who does not depart prior to the clearance void time must advise ATC as soon as possible of their intentions. This time cannot exceed 30 minutes. Failure to contact ATC within 30 minutes after the clearance void time will result in the aircraft being considered overdue and search and rescue procedures initiated.

Air Traffic Control

1) **Components of Air Traffic Control**.
   a) Air Route Traffic Control Centers (ARTCCs) have jurisdiction mostly over en route IFR flights. In the contiguous United States, there are twenty centers which control airspace. Jurisdiction boundaries are shown on en route charts.
      i) Each ARTCC coordinates traffic within its area, and alerts the next ARTCC or approach control of your specific information as you approach the area.
      ii) Each ARTCC is divided into sectors. These boundaries are not charted, since you are "handed off" to a controller in each new sector.
   b) Terminal Radar Control Facilities (TRACONs) regulate air traffic around terminal areas. TRACON systems normally extend to between 10,000 and 12,000 feet MSL, and as far away as 30 N.M. from the primary airport.
      i) Non-radar approach control usually extends between 3000 & 4000 AGL. IFR approaches and departures for airports outside of approach control areas are the jurisdiction of the ARTCC.

2) **Federal Airways** – Airways are typically eight nautical miles wide, and may classified by altitude.
    a) Airways based upon ground based navaids with a midpoint segment or a changeover point more than 51 N.M. from a navaid will be wider than 8 N.M. for the portion of the airway which exceeds this 51 N.M. limit. This additional width is based upon a 4.5° angle from the navaid on each side the airway centerline.
    b) Victor Airways are VOR airways used below 18,000 ft MSL.
    c) L/MF airways (colored airways) are predicated solely on L/MF navigation aids and are depicted in brown on aeronautical charts and are identified by color name and number.
    d) Jet Routes are VOR airways used from 18,000 ft MSL to and including FL 450.
    e) Published Area Navigation (RNAV) routes require RNAV capability. Above 18,000 ft. MSL, these routes are known as "Q" routes over the United States.
3) **Air Traffic Service (ATS) Routes**- The term "ATS route" is a generic term that includes "VOR Federal airways," "colored Federal airways," "alternate airways," "jet routes," "Military Training Routes," "named routes," and "RNAV routes." The term "ATS route" does not replace these more familiar route names, but serves only as an overall title when listing the types of routes that comprise the United States route structure.

Navaid Service Volumes

1) Low-altitude navaids are used for navigation below 18,000 feet MSL. Their service volume is 40 N.M., so that maximum distance between VOR stations on a direct flight is 80 N.M.
2) High-Altitude navaids are used for navigation for all altitude strata. Up to 14,500 feet MSL, the service volume is 40 N.M. From 14,500 feet MSL up to but not including 18,000 feet MSL, the service volume increases to 100 N.M. From 18,000 feet MSL to FL450, service volume increases to 130 N.M., while above FL 450 service volume decreases to 100 N.M.
3) MEA, or the Minimum En Route Altitude, is the lowest altitude between radio fixes, which assures satisfactory navigation signal coverage and minimum obstacle clearance requirements. The MEA applies to the total width of the airway.
4) MOCA, or the Minimum Obstruction Clearance Altitude, is the lowest altitude which assures clearance requirements over obstacles within the boundaries a VOR airway, an off-airway route, or a route segment centerline, and assures acceptable navigation signal coverage within 22 N.M. of a navaid.
5) OROCA, or the Off-Route Obstruction Clearance Altitude, provides obstruction clearance while traveling off of federal airways. This altitude may not provide signal coverage from ground-based navigational aids, air traffic control radar, or communications coverage.

En Route

1) Direct routes are flights between navaids, airports, or waypoints that do not have an airway between them, and have not been checked for VOR interference.
2) When changing altitude, controllers expect climb or descent to be made as rapidly as possible until within 1000 feet of the new assigned altitude. The last 1000 feet should be made between 500 and 1,500 fpm.
    a) The controller should be notified any time you are unable to:
        i) Maintain a climb or descent of at least 500 fpm
        ii) Maintain the cruising true airspeed specified on the flight plan within 5% or 10 knots, whichever is greater.

Flight Plans

1) According to 14 CFR §91.169, no person may operate an aircraft IFR in controlled airspace unless a flight plan has been filed.
    a) No clearance is required or available in uncontrolled airspace (class G).

Flight Plan Form

Form Approved: OMB No. 2120-0034

U.S. DEPARTMENT OF TRANSPORTATION
FEDERAL AVIATION ADMINISTRATION
FLIGHT PLAN

(FAA USE ONLY) ☐ PILOT BRIEFING ☐ VNR ☐ STOPOVER | TIME STARTED | SPECIALIST INITIALS

1. TYPE	2. AIRCRAFT IDENTIFICATION	3. AIRCRAFT TYPE/ SPECIAL EQUIPMENT	4. TRUE AIRSPEED	5. DEPARTURE POINT	6. DEPARTURE TIME		7. CRUISING ALTITUDE
VFR					PROPOSED (Z)	ACTUAL (Z)	
X IFR							
DVFR			KTS				

8. ROUTE OF FLIGHT

9. DESTINATION (Name of airport and city)	10. EST. TIME ENROUTE		11. REMARKS
	HOURS	MINUTES	

12. FUEL ON BOARD		13. ALTERNATE AIRPORT(S)	14. PILOT'S NAME, ADDRESS & TELEPHONE NUMBER & AIRCRAFT HOME BASE	15. NUMBER ABOARD
HOURS	MINUTES		17. DESTINATION CONTACT/TELEPHONE (OPTIONAL)	

16. COLOR OF AIRCRAFT

CIVIL AIRCRAFT PILOTS. FAR Part 91 requires you file an IFR flight plan to operate under instrument flight rules in controlled airspace. Failure to file could result in a civil penalty not to exceed $1,000 for each violation (Section 901 of the Federal Aviation Act of 1958, as amended). Filing of a VFR flight plan is recommended as a good operating practice. See also Part 99 for requirements concerning DVFR flight plans.

FAA Form 7233-1 (8-82) CLOSE VFR FLIGHT PLAN WITH ________________ FSS ON ARRIVAL

2) Items necessary on an IFR flight plan:
   a) **Type of Flight Plan** -- IFR, VFR, or DVFR.
   b) **Aircraft Identification Number**.
   c) **Aircraft Type** -- Transponder, RNAV, and DME information is also included in this item. Special equipment suffix letters are listed in the AIM, paragraph 5-1-7. The AIM chart has been reproduced on the following page.
   d) **True Airspeed** -- Reported in knots.
   e) **Departure Point**.
   f) **Departure Time** -- Estimated time should be filled out upon filing. Actual time will be recorded after departure.
   g) **Cruising Altitude** -- This is the requested en route altitude for the first leg of the trip, and therefore a different altitude may be assigned. File odd altitudes for eastbound flights (0° - 179°) magnetic and even altitudes for westbound flights (180° - 359°).
   h) **Route of Flight** -- By airways, navaids, or waypoints.
   i) **Destination Airport and City**.
   j) **Estimated Time En Route** -- This should be figured as the estimated time over the point of first intended landing.
   k) **Remarks** -- Enter only those pertinent to ATC if any.
   l) **Fuel on Board** -- Usable fuel available given in hours and minutes at the cruising airspeed listed in box 4.
   m) **Alternate Airport** -- For a precision approach procedure, the alternate must be forecast to have at the estimated time of arrival at least a 600 foot ceiling and 2 statute miles visibility; for a nonprecision approach, the alternate must be forecast to have at least an 800 foot ceiling and 2 statute miles of visibility. If the alternate has no instrument approach, VFR conditions must be forecast from the MEA to landing.
      i) NOTE: An alternate airport is not required if the first airport of intended landing has an instrument approach and is forecast to have at least a 2,000 foot ceiling and 3 miles of visibility for at least one hour before and after the estimated time of arrival.
   n) **Pilot's Full Name, Address, Telephone, and Aircraft Home Base**.
   o) **Number of Persons on Board**.
   p) **Color of Aircraft**.
   q) **Destination Contact and Telephone** – Optional.
3) Flight plans should be filed at least 30 minutes before the estimated time of departure (ETD). Flight plans are normally kept by ATC for at least one hour after ETD.

4) IFR flight plans are automatically closed upon landing when the destination airport has an operating control tower. For flights to airports without an operating control tower, the pilot is responsible for initiating cancellation of the flight plan. According to the AIM, paragraph 5-1-13, the flight plan can be closed either:

   a) After landing at an airport with an operating FSS, or,

   b) While still airborne and in contact with ATC. This method is preferable to closing the flight plan by telephone after landing, since it frees the airspace immediately for use by other aircraft.

Suffix	Equipment Capability
	NO DME
/X	No transponder
/T	Transponder with no Mode C
/U	Transponder with Mode C
	DME
/D	No transponder
/B	Transponder with no Mode C
/A	Transponder with Mode C
	TACAN ONLY
/M	No transponder
/N	Transponder with no Mode C
/P	Transponder with Mode C
	AREA NAVIGATION (RNAV)
/Y	LORAN, VOR/DME, or INS with no transponder
/C	LORAN, VOR/DME, or INS, transponder with no Mode C
/I	LORAN, VOR/DME, or INS, transponder with Mode C
	ADVANCED RNAV WITH TRANSPONDER AND MODE C (If an aircraft is unable to operate with a transponder and/or Mode C, it will revert to the appropriate code listed above under Area Navigation.)
/E	Flight Management System (FMS) with DME/DME and IRU position updating
/F	FMS with DME/DME position updating
/G	Global Navigation Satellite System (GNSS), including GPS and Wide Area Augmentation System (WAAS), with en route and terminal capability
/R	Required Navigational Performance (RNP). The aircraft meets the RNP type prescribed for the route segment(s), route(s) and/or area concerned.
	Reduced Vertical Separation Minimum (RVSM). Prior to conducting RVSM operations within the U.S., the operator must obtain authorization from the FAA or from the responsible authority, as appropriate.
/J	/E with RVSM
/K	/F with RVSM
/L	/G with RVSM
/Q	/R with RVSM
/W	RVSM

Volume 2 – Questions 3.5 & 5.8

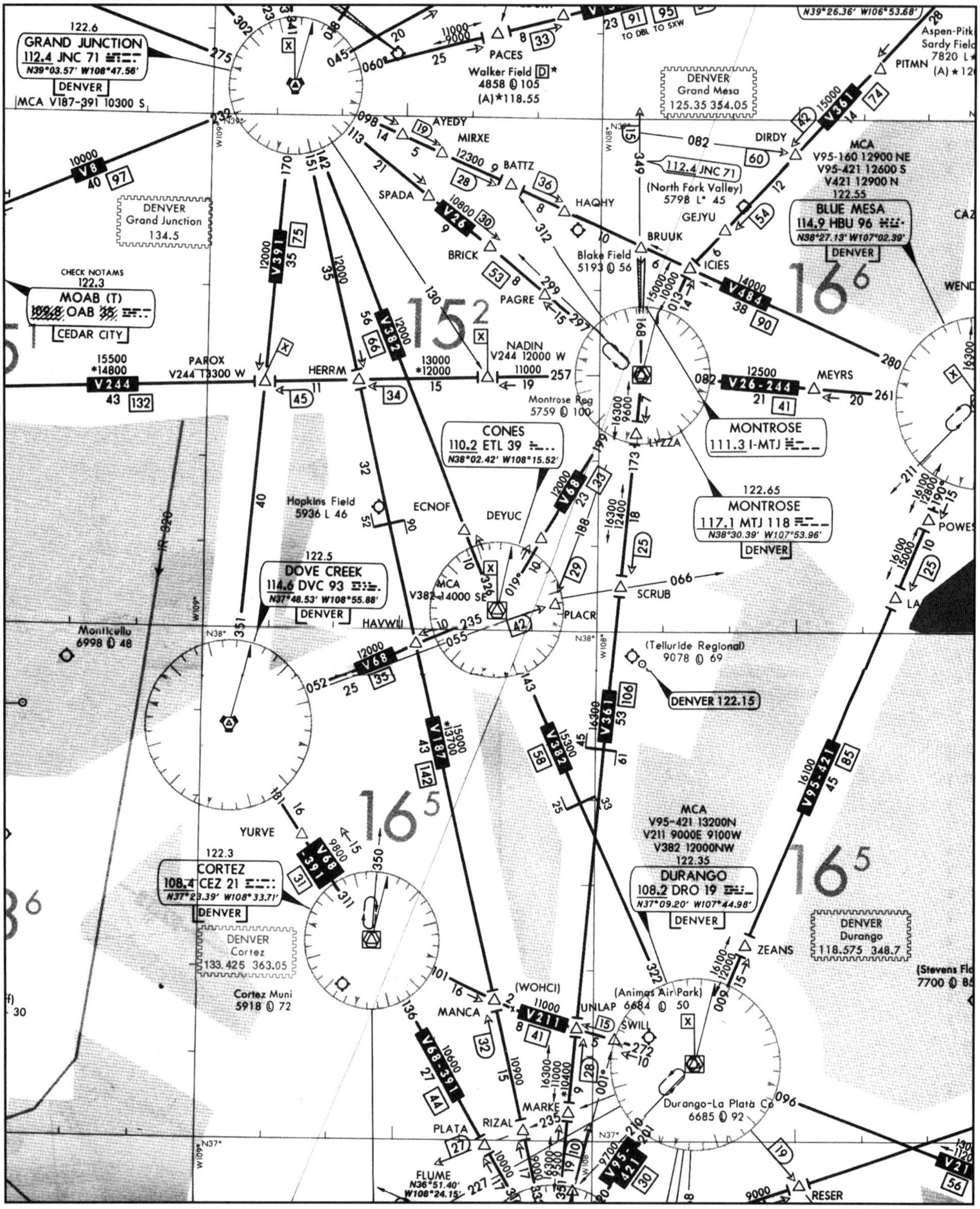

Volume 3 - Instrument Approaches

The goal of an instrument approach is to transition an IFR aircraft from the en route system to a point where the airport runway is in sight, where it can complete the landing procedure visually. IFR approaches are categorized as **precision, nonprecision,** or an **approach with vertical guidance**.

Approach & Landing Charts

1) Approach charts are very detailed, and give all information needed for the approach procedures. Each approach chart is broken down into multiple sections:
   a) The **Pilot Briefing Information** section is used when preparing for the approach. It contains a summary of the information needed for the approach. It includes NAVAID, final approach course, and airport information as appropriate; a **Remarks** section with any special advisories of which the pilot should be aware; a textual version of the **Missed Approach** instructions; and a listing of the pertinent communication frequencies.
   b) The **Planview** is used for the initial approach segment and gives a bird's eye view of the approach.
   c) The **Profile** view is used after the approach has started. It gives information concerning altitudes at different locations on the approach and includes an iconic representation of the **Missed Approach** instructions.
   d) The **Minimums** section is used near the end of an approach, and contains information about the minimum flight visibility and minimum altitude that can be achieved before a decision is made concerning the success of the approach.
   e) The **Airport Sketch** gives a view of all runways and taxiways on the airport.

Precision Approaches

1) Precision approaches are flown at an angle of descent. The purpose of an instrument approach is to bring an aircraft down through clouds to a point where the runway is visible. If the runway is not visible at a certain altitude on a precision approach, a missed approach must be flown. This altitude is called the **Decision Altitude (DA)**. At this MSL altitude on the glide slope, the decision must be made either to continue the approach or to fly a missed approach. **Decision Height (DH)** is the height above the threshold elevation where this decision must be made.
2) Decision Altitude (DA) is referenced to mean sea level (MSL) and Decision Height (DH) is referenced to the threshold elevation.
3) Precision approaches are standard instrument approaches with glide slope information meeting the precision standards of ICAO Annex 10; for example, the Instrument Landing System (ILS), Microwave Landing System (MLS), and Precision Approach Radar (PAR). ILS is by far the most common; MLS and PAR are rare.
   a) An approach that provides glidepath deviation information but does not meet the precision requirements of ICAO Annex 10, is termed an Approach with Vertical Guidance (APV).
4) Guidance information for ILS is provided by the localizer for left/right guidance, and the glide slope for up/down guidance.
5) **The Localizer**.
   a) The Localizer is an electronic extension of the centerline of the runway, and operates within the frequency range of 108.10 to 111.95. Their three-letter identification codes are always preceded by the letter "I" for easy identification. Localizers are adjusted so that the signal is 700 feet wide at the approach end of the runway. Due to varying runway lengths, the width of the localizer signal varies between 3° and 6°.
      i) A Localizer Directional Aid, or LDA, is a localizer that is not aligned with the centerline of the runway. An LDA is normally not a precision approach.
      ii) A Simplified Directional Facility, or SDF, is much like an LDA, but is less precise, measuring either 6° or 12° wide. SDF identifiers are not preceded by the letter "I". Glide slopes are not installed with SDFs. An SDF is NOT a precision approach.
   b) Heading corrections should be 5° or less unless you have a full-scale deflection. Also, as you approach the runway, the localizer gets smaller. Therefore, corrections should be proportionally smaller.
   c) Localizers normally emit signals in two directions, called the Front Course and the Back Course. The front course is used for the ILS approach. The back course is only usable when an approach procedure has been approved for it. The back course is marked on the chart in large bold letters. The back course inbound instrument indications are opposite of those for a front course approach.

6) **The Glide Slope.**

a) The Glide Slope provides vertical navigation. Glide slope frequencies are paired with localizer frequencies, and most receivers are designed to automatically select the appropriate glide slope frequency when a localizer frequency is selected. Glide slopes are positioned to give a 3°-descent angle (318 feet per nautical mile), and usually have a threshold crossing height of 55 feet. The glide slope is extremely precise, and a full scale deflection either above or below is equal to a 0.7° or greater deviation from the centerline.

b) With an increase in ground speed, the descent rate must be increased in order to maintain the glide slope. Conversely, with a decrease in ground speed, the descent rate must be decreased. The table below can be used to compute the descent per minute given an approximate ground speed and angle of descent.

c) NOTE: SINCE GLIDE SLOPE INDICATIONS CAN BE RECEIVED ANYWHERE AROUND THE ANTENNA, THE GLIDE SLOPE SHOULD ONLY BE USED WHEN THE APPROACH CHART SPECIFIES A GLIDE SLOPE.

Rate of Descent Table

DESCENT TABLE 99028

RATE OF DESCENT TABLE

A rate of descent table is provided for use in planning and executing precision descents under known or approximate ground speed conditions. It will be especially useful for approaches when the localizer only is used for course guidance. A best speed, power, altitude combination can be programmed which will result in a stable glide rate and altitude favorable for executing a landing if minimums exist upon breakout. Care should always be exercised so that minimum descent altitude and missed approach point are not exceeded.

	ANGLE OF DESCENT (degrees and tenths)	FEET /NM	GROUND SPEED (knots)										
			30	45	60	75	90	105	120	135	150	165	180
	2.0	210	105	160	210	265	320	370	425	475	530	585	635
	2.5	265	130	200	265	330	395	465	530	595	665	730	795
VERTICAL PATH ANGLE	2.7	287	143	215	287	358	430	501	573	645	716	788	860
	2.8	297	149	223	297	371	446	520	594	669	743	817	891
	2.9	308	154	231	308	385	462	539	616	693	769	846	923
	3.0	318	159	239	318	398	478	557	637	716	796	876	955
	3.1	329	165	247	329	411	494	576	658	740	823	905	987
	3.2	340	170	255	340	425	510	594	679	764	849	934	1019
	3.3	350	175	263	350	438	526	613	701	788	876	963	1051
	3.4	361	180	271	361	451	541	632	722	812	902	993	1083
	3.5	370	185	280	370	465	555	650	740	835	925	1020	1110
	4.0	425	210	315	425	530	635	740	845	955	1060	1165	1270
	4.5	475	240	355	475	595	715	835	955	1075	1190	1310	1430
	5.0	530	265	395	530	660	795	925	1060	1190	1325	1455	1590
	5.5	580	290	435	580	730	875	1020	1165	1310	1455	1600	1745
	6.0	635	315	475	635	795	955	1110	1270	1430	1590	1745	1950
	6.5	690	345	515	690	860	1030	1205	1375	1550	1720	1890	2065
	7.0	740	370	555	740	925	1110	1295	1480	1665	1850	2035	2220
	7.5	795	395	595	795	990	1190	1390	1585	1785	1985	2180	2380
	8.0	845	425	635	845	1055	1270	1480	1690	1905	2115	2325	2540
	8.5	900	450	675	900	1120	1345	1570	1795	2020	2245	2470	2695
	9.0	950	475	715	950	1190	1425	1665	1900	2140	2375	2615	2855
	9.5	1005	500	750	1005	1255	1505	1755	2005	2255	2510	2760	3010
	10.0	1055	530	790	1055	1320	1585	1845	2110	2375	2640	2900	3165
	10.5	1105	555	830	1105	1385	1660	1940	2215	2490	2770	3045	3320
	11.0	1160	580	870	1160	1450	1740	2030	2320	2610	2900	3190	3480
	11.5	1210	605	910	1210	1515	1820	2120	2425	2725	3030	3335	3635
	12.0	1260	630	945	1260	1575	1890	2205	2520	2835	3150	3465	3780

DESCENT TABLE 99028

7) **Range information** is provided by marker beacons. Marker beacons transmit a signal straight up. All marker beacons transmit on 75 megahertz.

a) Outer Markers are normally located between 4 and 7 miles from the approach end of the runway. It is normally located at or near the glide slope intercept altitude. The outer marker transmits dashes at a rate of 2 per second, and the three light marker indicator illuminates a blue light.

b) Middle Markers are located about 3500 feet from the approach end of the runway. It is normally located at or near the decision height along the glide slope. The middle marker transmits alternating dots and dashes at a rate of 95 pairs per second, and the middle marker illuminates the amber light.

8) MLS approaches are extremely rare. For a description of MLS operation, consult the Aeronautical Information Manual.

Runway Markings - Used on Runways Served by Precision Instrument Approaches

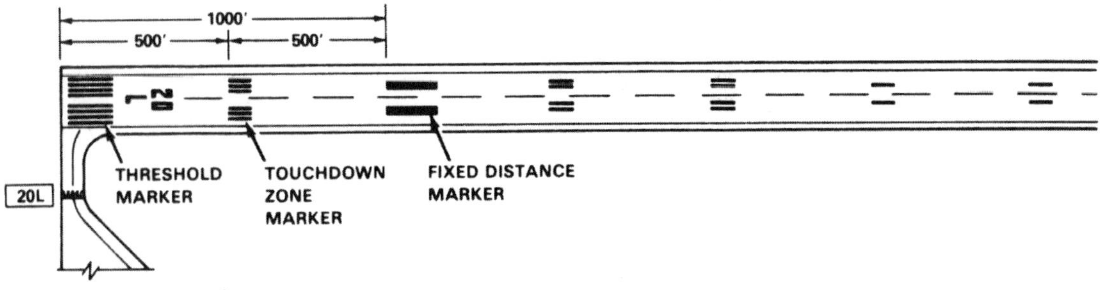

Nonprecision Approaches

1) Nonprecision approaches include GPS, VOR, NDB, TACAN, and SDF approaches. Nonprecision approaches do not provide glide slope information. Therefore, instead of a Decision Altitude, a nonprecision approach has a Minimum Descent Altitude, or MDA. This altitude is flown in level flight until the airport is in sight or until a missed approach is begun.

Approaches with Vertical Guidance (APV)

1) APV approaches provide lateral and vertical guidance but does not meet ICAO Annex 10 standards required for precision approaches. These include Lateral Navigation / Vertical Navigation (LNAV/VNAV), Localizer Performance with Vertical guidance (LPV), and LDA with glide slope

Runway Lighting Systems

1) Touchdown Zone Lights are placed on either edge of the runway at the approach end. The system normally extends 3,000 feet along the runway, and lights are placed at 100-foot intervals.

2) Runway Centerline Lighting begins 75 feet from the landing threshold and ends 75 feet from the opposite end of the runway. They are spaced 50 feet apart.

Circling Approaches

1) A Circling Approach is a maneuver initiated by the pilot to align the aircraft with the runway for landing when a straight-in landing from an instrument approach is not possible. In order to fly a circling approach, the pilot must have ATC authorization and must have established the required visual reference to the airport. The minimums for a circling approach are listed in the minimums section of the approach chart. Navigation is visual.

Radar Approaches

1) A radar approach is an approach where the air traffic controller issues instructions, for pilot compliance, based on aircraft position in relation to the final approach course and the distance from the end of the runway as displayed on the controller's radar display. These may be given to any aircraft upon request and may be offered to pilots of aircraft in distress or to expedite traffic. There are two types of radar approaches. Each type may be performed at airports where minimums are published for the particular type of approach.

2) A Precision Approach Radar (PAR) approach is one where the controller provides highly accurate navigational guidance for runway alignment and elevation. Pilots will be given headings to fly, and are told when to expect glide path interception and when to begin descent. This information is provided until the aircraft reaches the published decision height, and radar service is automatically terminated upon completion of the approach.

3) An Airport Surveillance Radar (ASR) approach is one in which a controller provides navigational guidance only, with no controller service for elevation. The pilot is furnished headings for runway centerline alignment as well as range information. Information pertaining to the commencement of descent to the MDA and when the aircraft is at the missed approach point is furnished as well. If requested by the pilot, the controller will furnish recommended altitudes each mile from the runway.

Approach Standards

1) The primary area extends beyond the 10-mile procedure turn limit, and the minimum obstacle clearance is 1,000 feet. It is also 1,000 feet for the entry area and maneuvering zone. Obstruction clearance in the secondary area starts at 500 feet and tapers to zero. Minimum obstacle clearance in circling areas is 300 feet, while a straight-in final approach segment may have a clearance as low as 250 feet.
2) The optimum descent gradient is 250 feet per mile with a maximum of 500 feet per mile during the initial segment.
3) Procedure turns can be reduced to five miles for category A aircraft, and extended to 15 miles for category E aircraft.

Procedure Turns

1) Procedure turns are prescribed when it is necessary to reverse direction to establish the approaching aircraft inbound on an intermediate or final approach course. Procedure turn directions are shown on charts, and must be performed in the direction shown by the chart for obstacle clearance purposes. The procedure turn or hold in lieu of procedure turn is a required maneuver when it is necessary to perform a course reversal. The procedure turn is not required when the symbol "No PT" is shown, when RADAR VECTORING to the final approach course is provided, when conducting a timed approach, or when the procedure turn is not authorized. The hold in lieu of procedure turn is not required when RADAR VECTORING to the final approach course is provided or when "No PT" is shown. The altitude prescribed for the procedure turn is a MINIMUM altitude until the aircraft is established on the inbound course. The maneuver must be completed within the distance specified in the profile view.
2) On FAA charts, a barbed arrow indicates the direction or side of the inbound course on which the procedure turn is made. Headings are provided for course reversal using the 45-degree type procedure turn, though this method is not required. The point at which the turn may be commenced and the type and rate of turn is left up to the pilot. Some procedure turns are specified by procedural track and MUST BE flown exactly as depicted.

Timed Approaches

1) Timed approaches are approaches initiated by ATC where successive aircraft are assigned a time to cross the approach fix inbound. They are indicated by ATC by assigning a time to depart the final approach fix inbound (nonprecision approach), or the marker or fix used in lieu of an outer maker (precision approach). Timed approaches may only be conducted under the following conditions:
    a) A control tower is in operation at the airport where the approaches are conducted.
    b) Direct communications are maintained between the pilot and the center or approach controller until the pilot is instructed to contact the tower.
    c) If more than one missed approach procedure, none require a course reversal.
    d) If only one missed approach procedure is available,
        i) Course reversal is not required, and
        ii) Reported ceiling and visibility are equal to or greater than the highest prescribed circling minimums for the IAP.
    e) Pilots shall not execute a procedure turn when cleared for the approach. It is the pilot's responsibility to adjust the holding pattern in order to cross the fix at the prescribed time.

Missed Approaches

1) A missed approach is a maneuver conducted by a pilot when an instrument approach cannot be completed to a landing. The route of flight and altitude are shown on instrument approach procedure charts.
2) Missed approach procedures must be initiated only at the missed approach point, and the pilot must comply with the procedures listed. Obstacle clearance areas assume that the missed approach procedure is started from the missed approach point at a point not lower than the MDA or DA.
3) A missed approach from a circling approach must first begin with a climbing turn towards the landing runway for obstacle avoidance, then roll out onto the heading given for the missed approach procedure.

Holding

1) A holding pattern is used whenever an aircraft is cleared to a fix other than the destination airport and delay is expected. If a holding pattern is charted and the controller does not issue complete holding instructions, the pilot is expected to hold as depicted on the appropriate chart. If it is not charted, or if the controller wishes a holding pattern other than that published to be flown, the controller will issue complete holding instructions.
2) An ATC clearance requiring an aircraft to hold where the pattern is not charted will include the following information:
    a) Direction of holding from the fix in terms of the eight cardinal compass points (N, NE, etc.)
    b) Holding fix
    c) Radial, course, bearing, airway or route on which the aircraft is to hold
    d) Leg length in miles if DME or RNAV is used. (Leg length will be specified in minutes on pilot request or if the controller considers it necessary).
    e) Direction of turn if either:
        i) Left turns are to be made
        ii) Clarification is requested by the pilot
        iii) Clarification is deemed necessary by the controller
    f) Time to expect further clearance and other delay information
3) Maximum holding pattern speed limits:

ALTITUDE (MSL)	AIRSPEED (KIAS)
MHA-6,000’	200
6,000’-14,000’	230
14,000’ and above	265

Volume 3 – Riverton VOR Runway 10

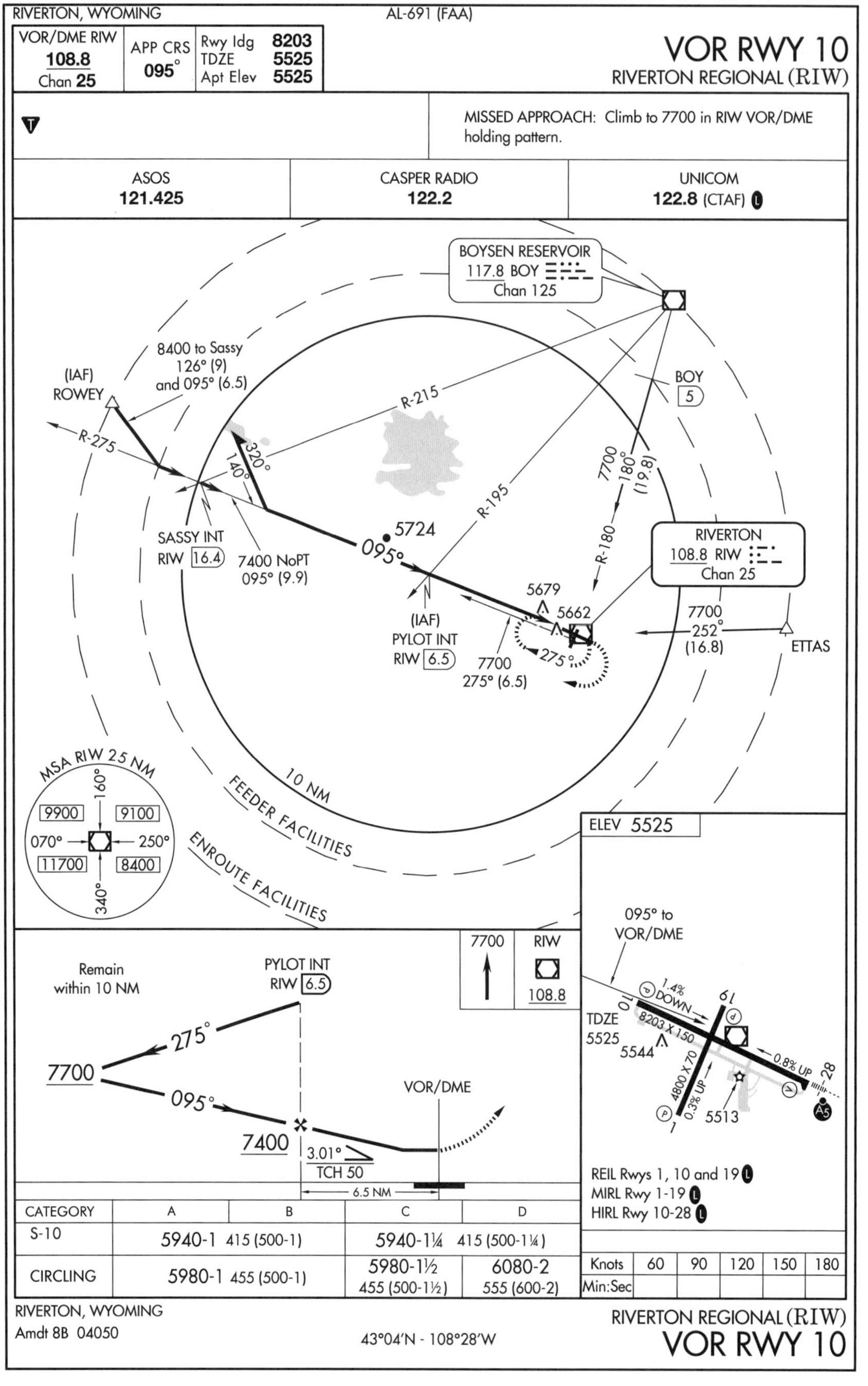

CATEGORY	A	B	C	D
S-10	5940-1 415 (500-1)		5940-1¼ 415 (500-1¼)	
CIRCLING	5980-1 455 (500-1)		5980-1½ 455 (500-1½)	6080-2 555 (600-2)

Volume 3 – Belfast NDB Runway 15

BELFAST, MAINE — AL-6448 (FAA)

NDB BST	APP CRS	Rwy ldg	4000
278	**144°**	TDZE	**195**
		Apt Elev	**195**

NDB RWY 15
BELFAST MUNI (BST)

T A NA — Use Bangor altimeter setting.	MISSED APPROACH: Climbing right turn to 2700 via BST 324° bearing to BOOGY Int and hold.
BANGOR APP CON **125.3 239.3**	UNICOM **122.8** (CTAF) L

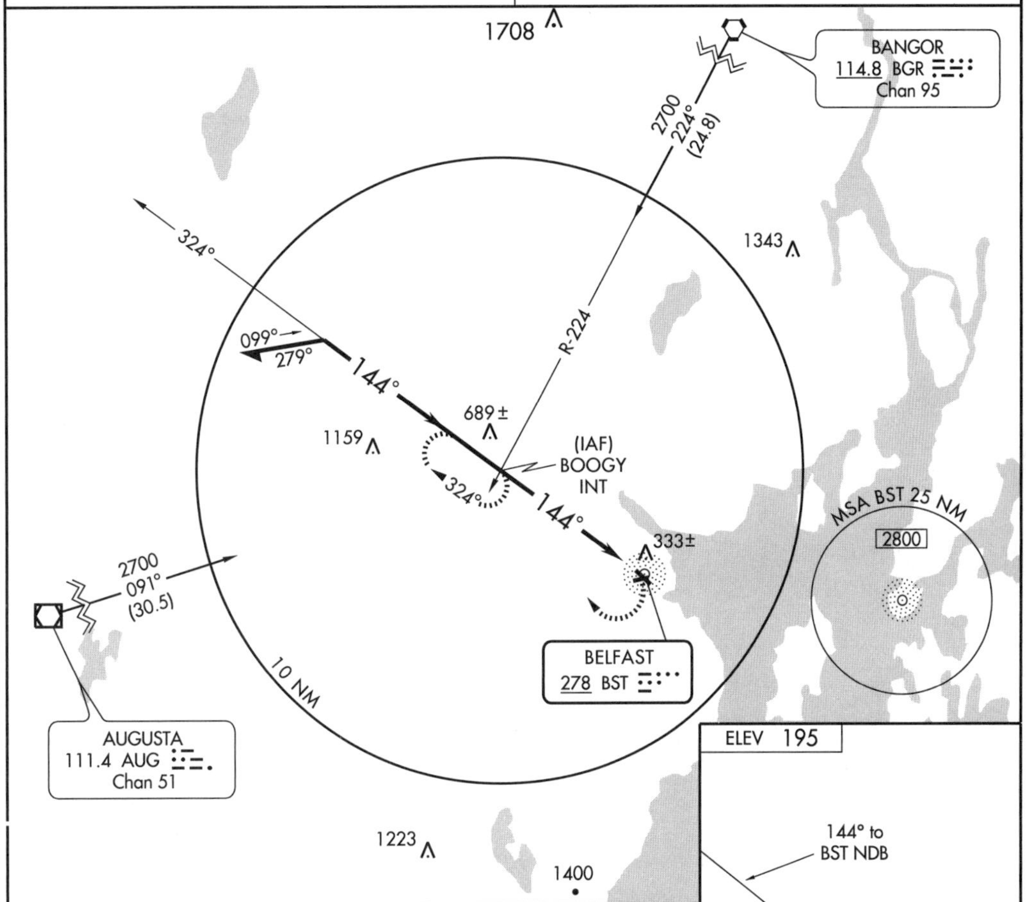

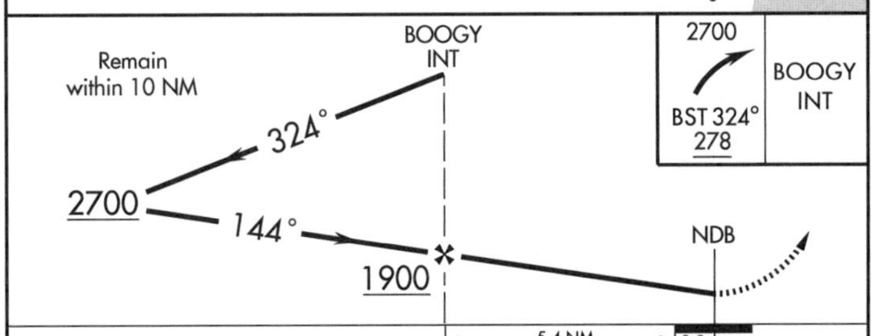

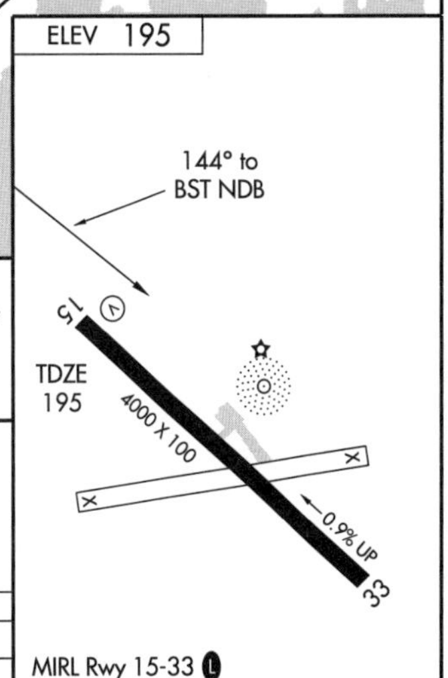

CATEGORY	A	B	C	D
S-15	840-1 645 (700-1)		840-1¾ 645 (700-1¾)	NA
CIRCLING	840-1 645 (700-1)		880-2 685 (700-2)	NA

Knots	60	90	120	150	180
Min:Sec					

BELFAST, MAINE
Amdt 3 05076

44°25'N - 69°01'W

BELFAST MUNI (BST)
NDB RWY 15

Volume 3 – Question 3.2

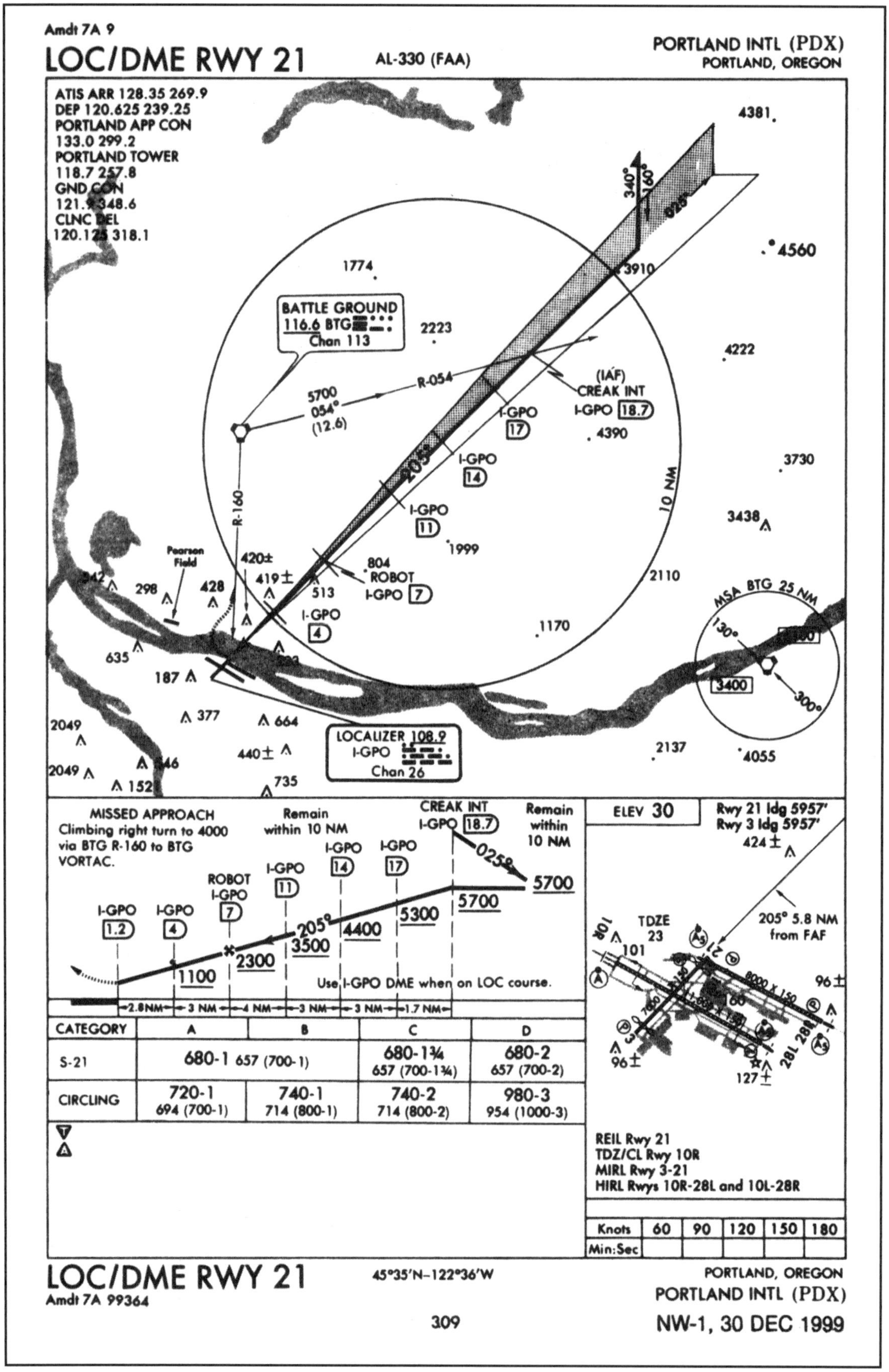

Volume 3 – Question 3.3

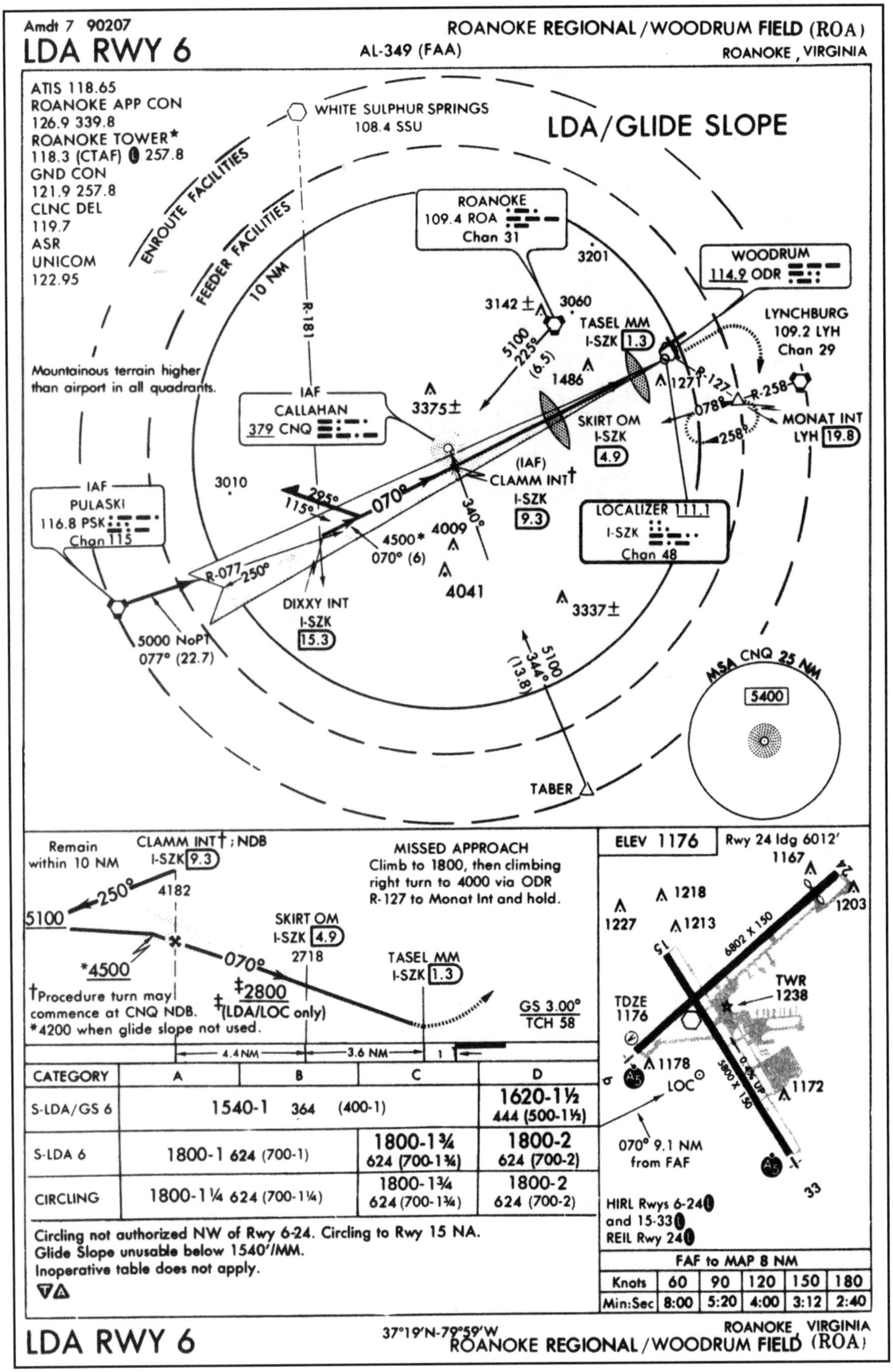

Volume 3 – Question 3.4

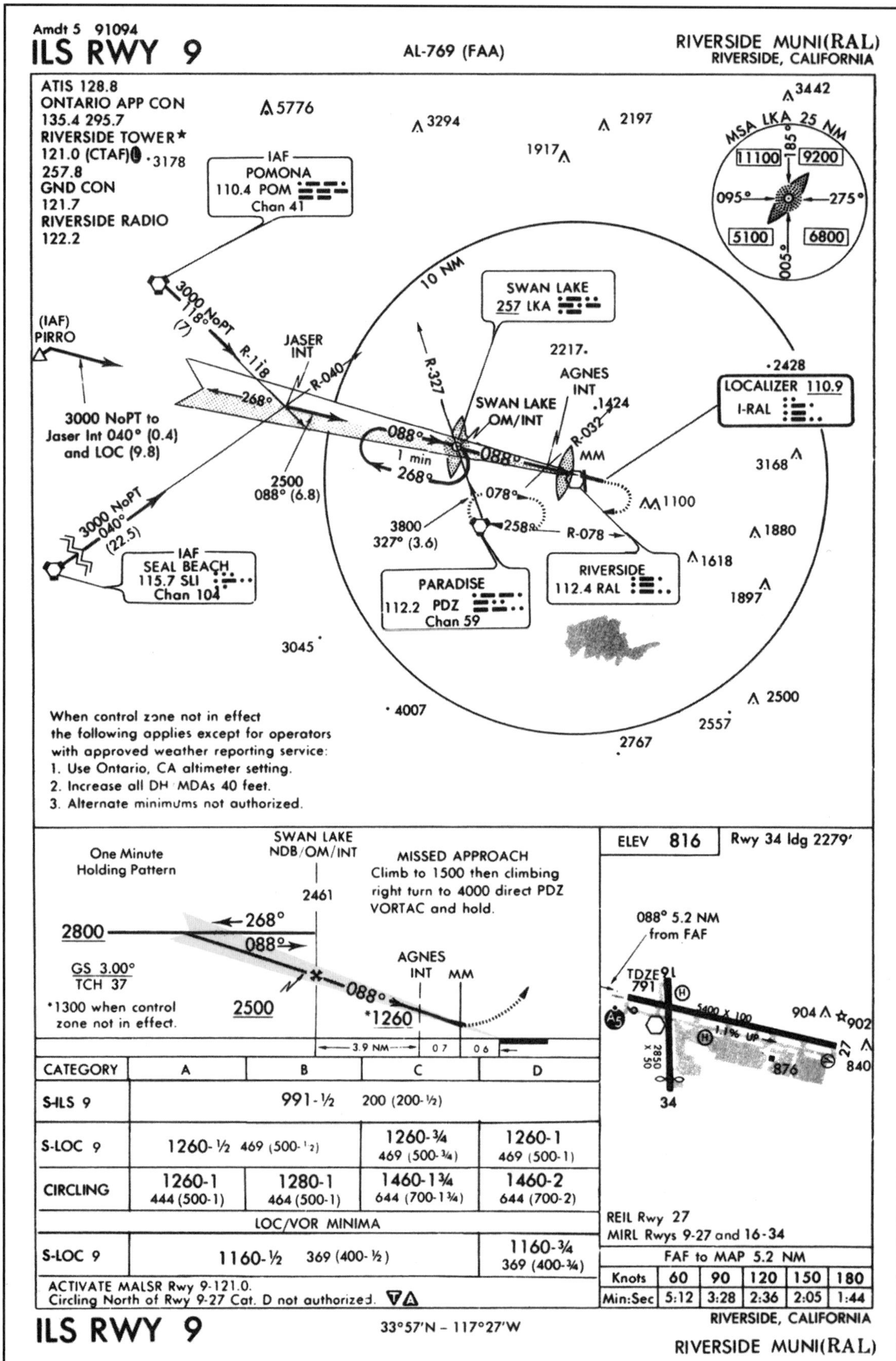

CATEGORY	A	B	C	D
S-ILS 9	991-½ 200 (200-½)			
S-LOC 9	1260-½ 469 (500-½)		1260-¾ 469 (500-¾)	1260-1 469 (500-1)
CIRCLING	1260-1 444 (500-1)	1280-1 464 (500-1)	1460-1¾ 644 (700-1¾)	1460-2 644 (700-2)
LOC/VOR MINIMA				
S-LOC 9	1160-½ 369 (400-½)			1160-¾ 369 (400-¾)

FAF to MAP 5.2 NM					
Knots	60	90	120	150	180
Min:Sec	5:12	3:28	2:36	2:05	1:44

Volume 3 – Question 3.5

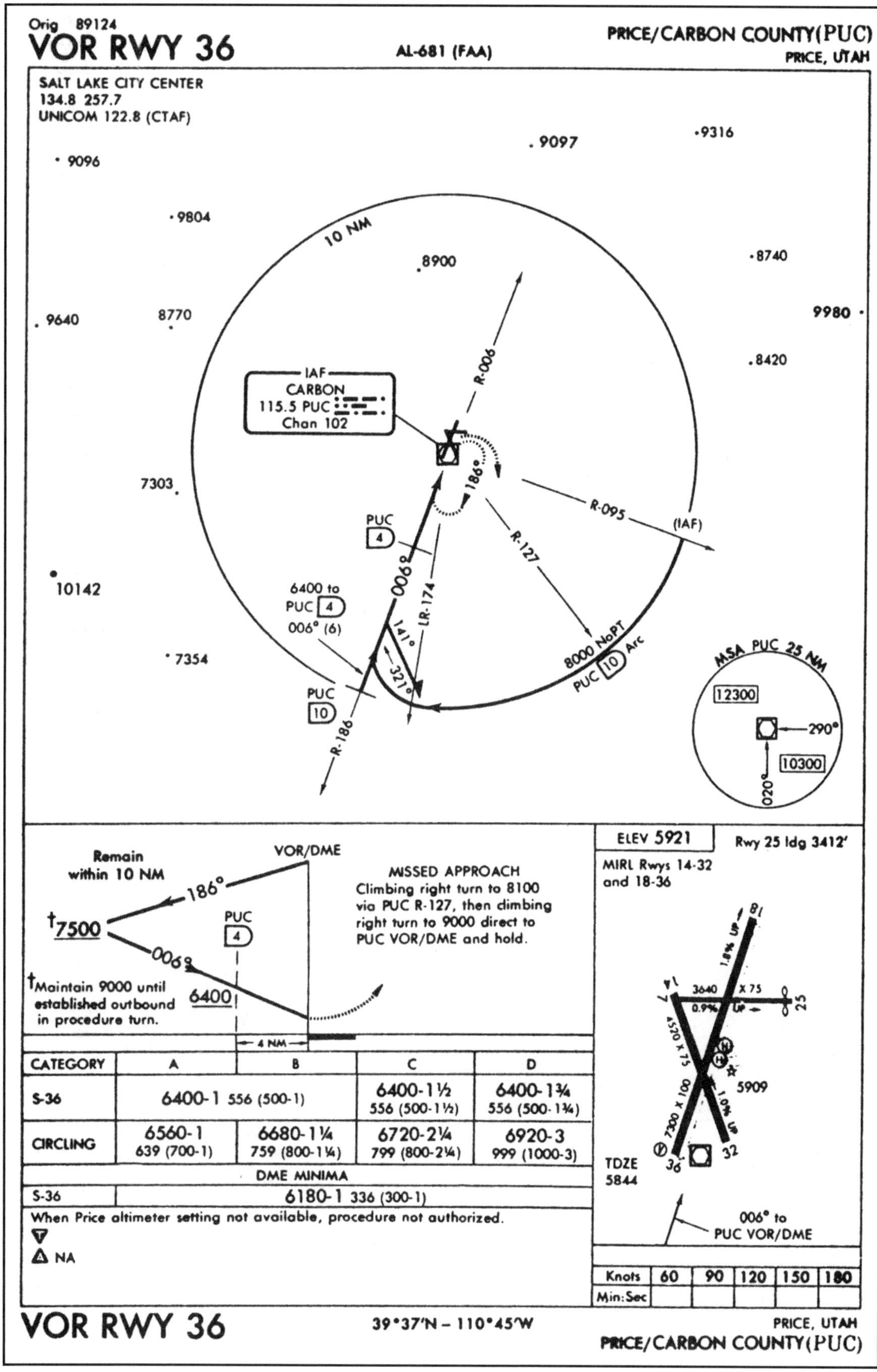

Volume 3 – Question 3.6

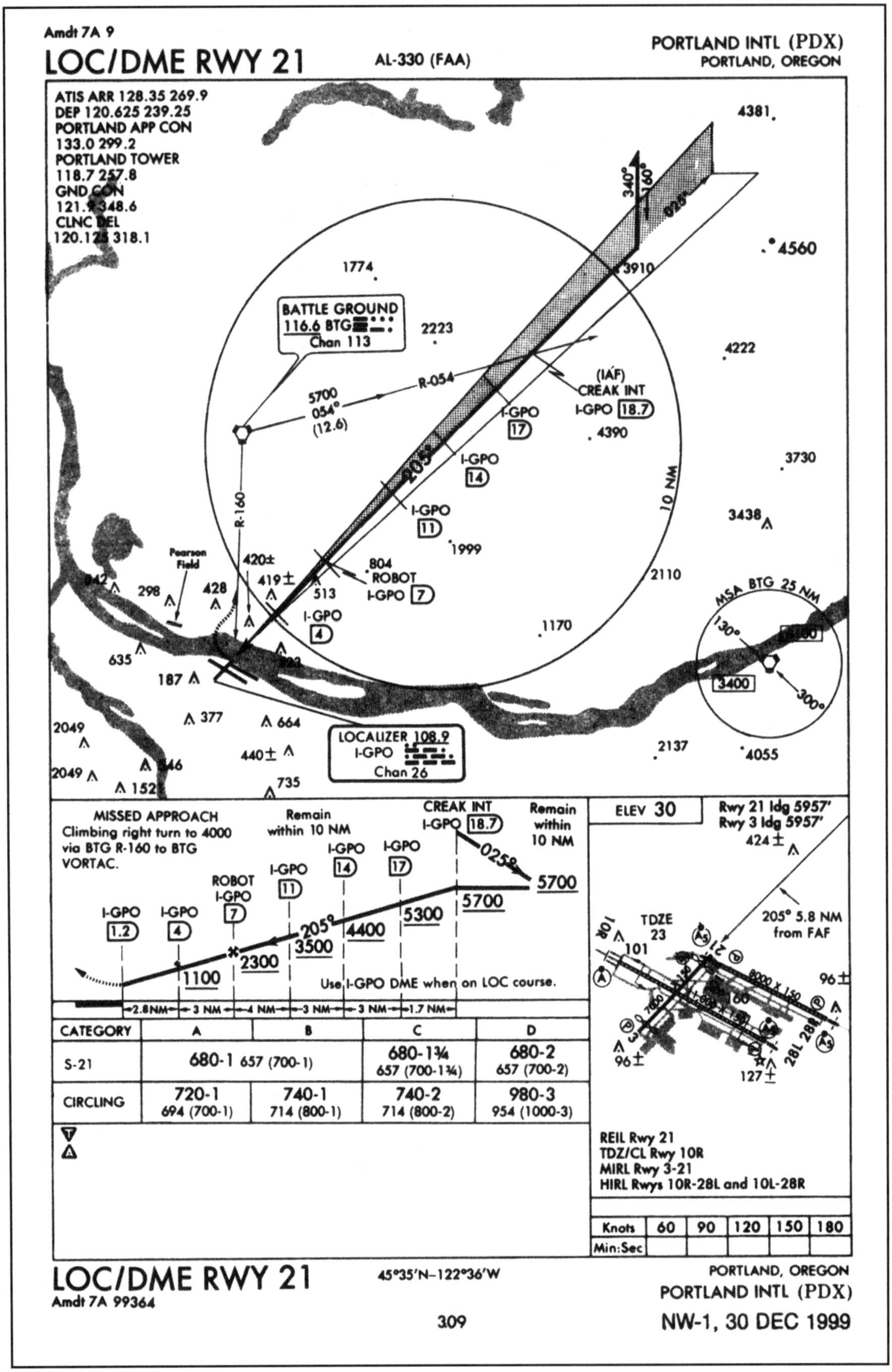

CATEGORY	A	B	C	D
S-21	680-1 657 (700-1)		680-1¾ 657 (700-1¾)	680-2 657 (700-2)
CIRCLING	720-1 694 (700-1)	740-1 714 (800-1)	740-2 714 (800-2)	980-3 954 (1000-3)

Knots	60	90	120	150	180
Min:Sec					

Volume 3 – Question 5.6

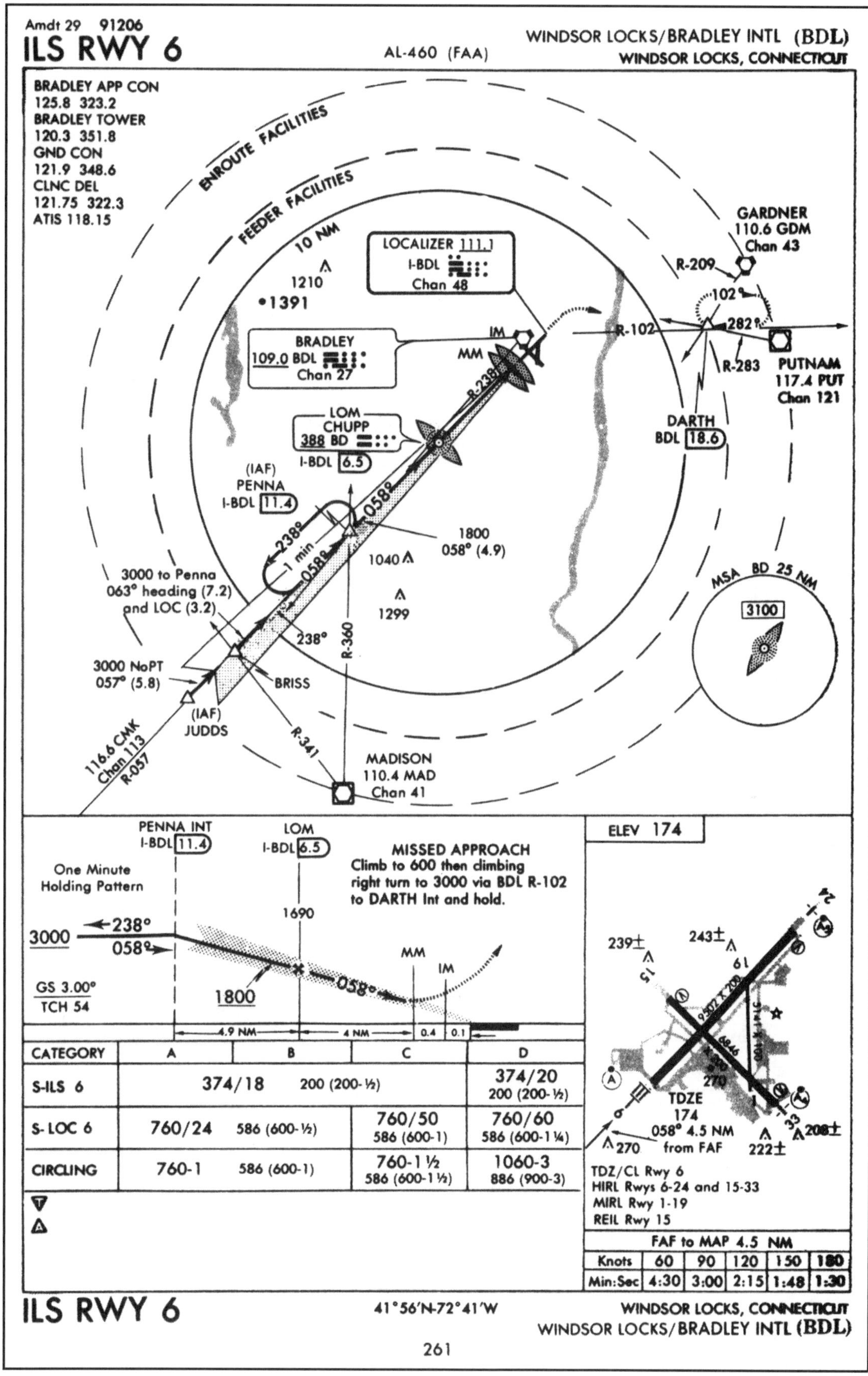

Volume 3 – Question 5.7

Amdt 21 99084

ILS RWY 24R

AL-237 (FAA)

LOS ANGELES INTL (LAX)
LOS ANGELES, CALIFORNIA

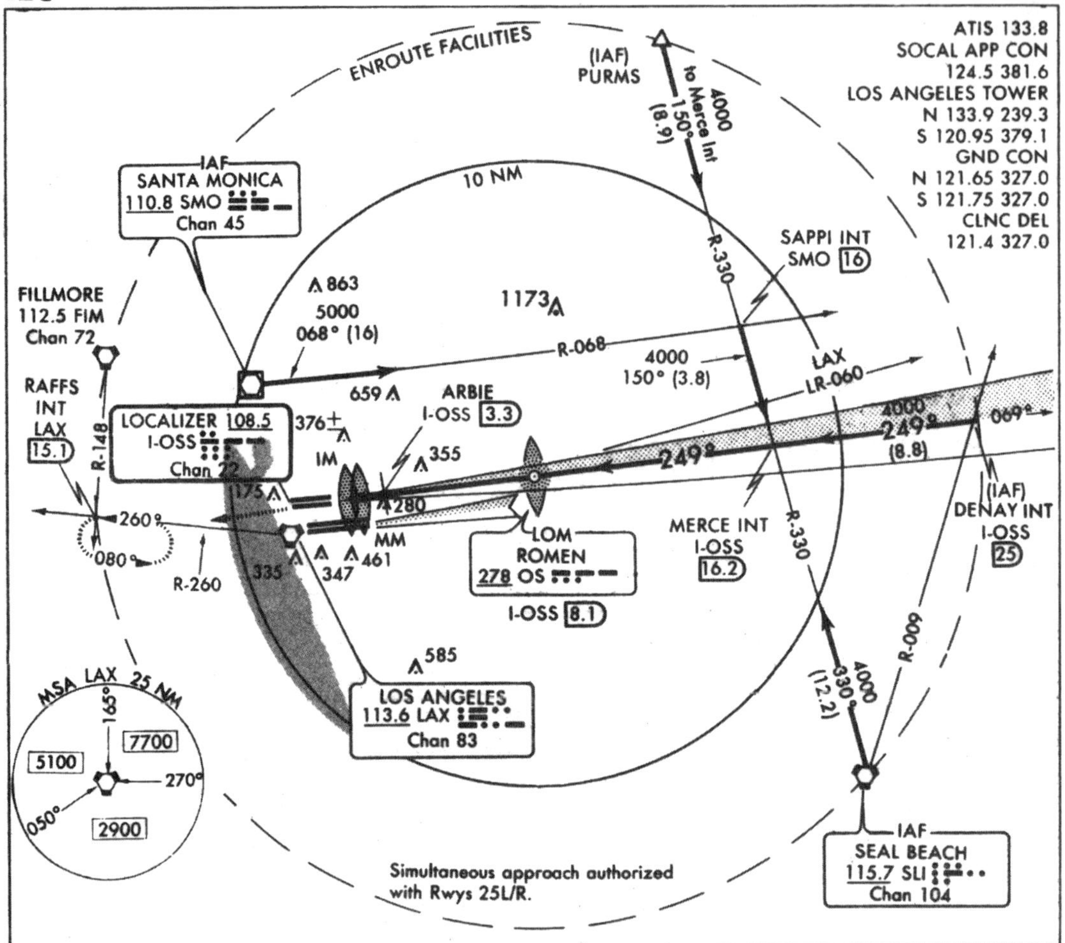

SW-2, 17 MAY 2001

MISSED APPROACH
Climb to 2000 via heading 249° and LAX R-260 to RAFFS Int/LAX 15.1 DME and hold.

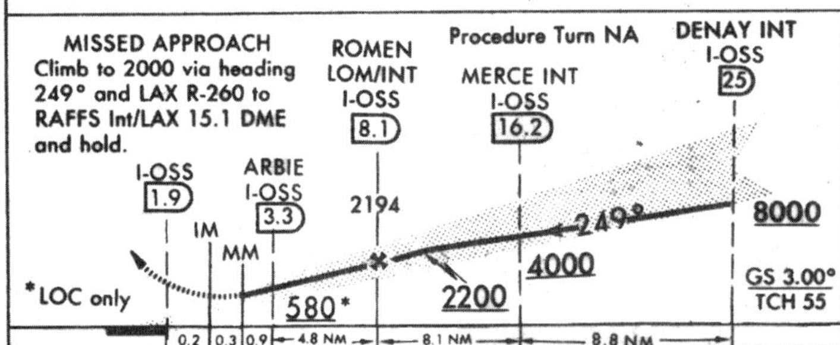

ELEV 126

Rwy 6R ldg 9954′
Rwy 25R ldg 11,134′

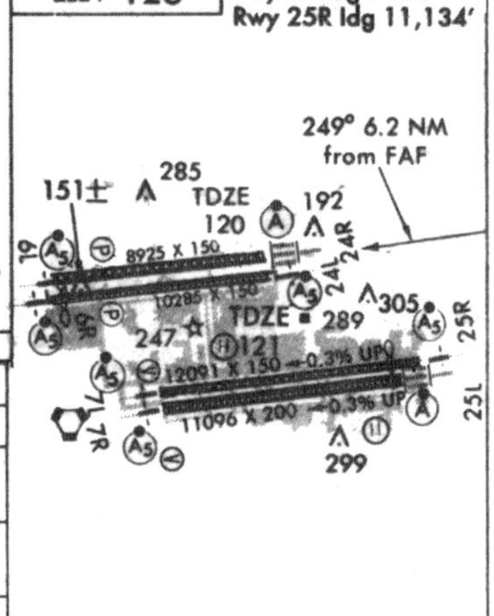

CATEGORY	A	B	C	D
S-ILS 24R	320/18 200 (200-⅜)			
S-LOC 24R	580/24 460 (500-½)		580/40 460 (500-¾)	580/50 460 (500-1)
SIDESTEP RWY 24L••	660/50 539 (600-1)			660-1½ 539 (600-1½)
ARBIE DME MINIMUMS				
S-LOC 24R†	460/24 340 (400-½)			460/40 340 (400-¾)

••Inoperative table does not apply to Cat. A and B.
†Cat. D visibility increased to RVR 5000 for inoperative ALSF-2.

TDZ/CL Rwys 6R, 7L, 24R and 25L
HIRL all Rwys

FAF to MAP 6.2 NM					
Knots	60	90	120	150	180
Min:Sec	6:12	4:08	3:06	2:29	2:04

ILS RWY 24R

Amdt 21 99084

33°57′N – 118°24′W

LOS ANGELES, CALIFORNIA
LOS ANGELES INTL (LAX)

177

Volume 3 – Question 9.1

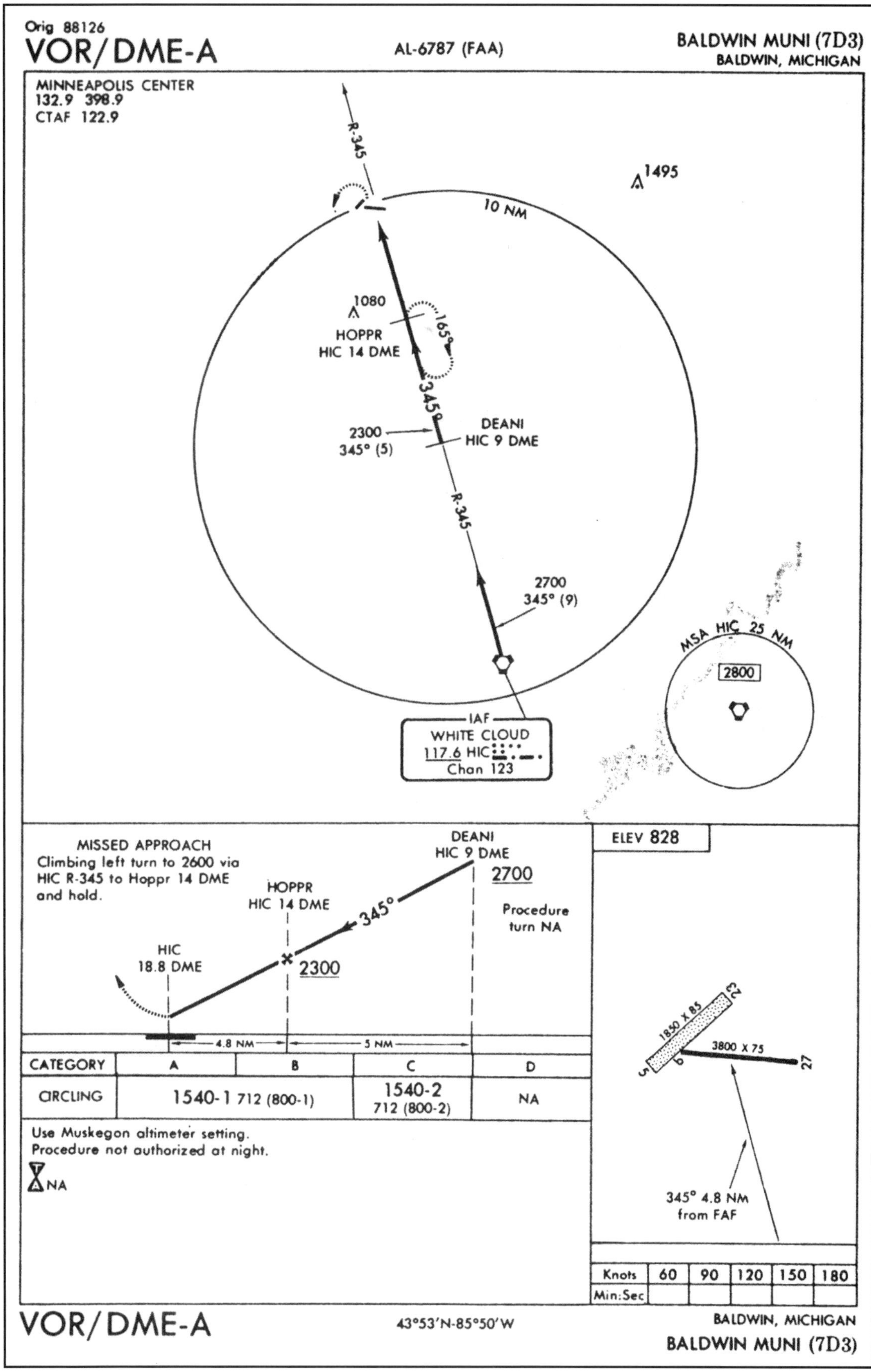

Volume 3 – Question 9.2

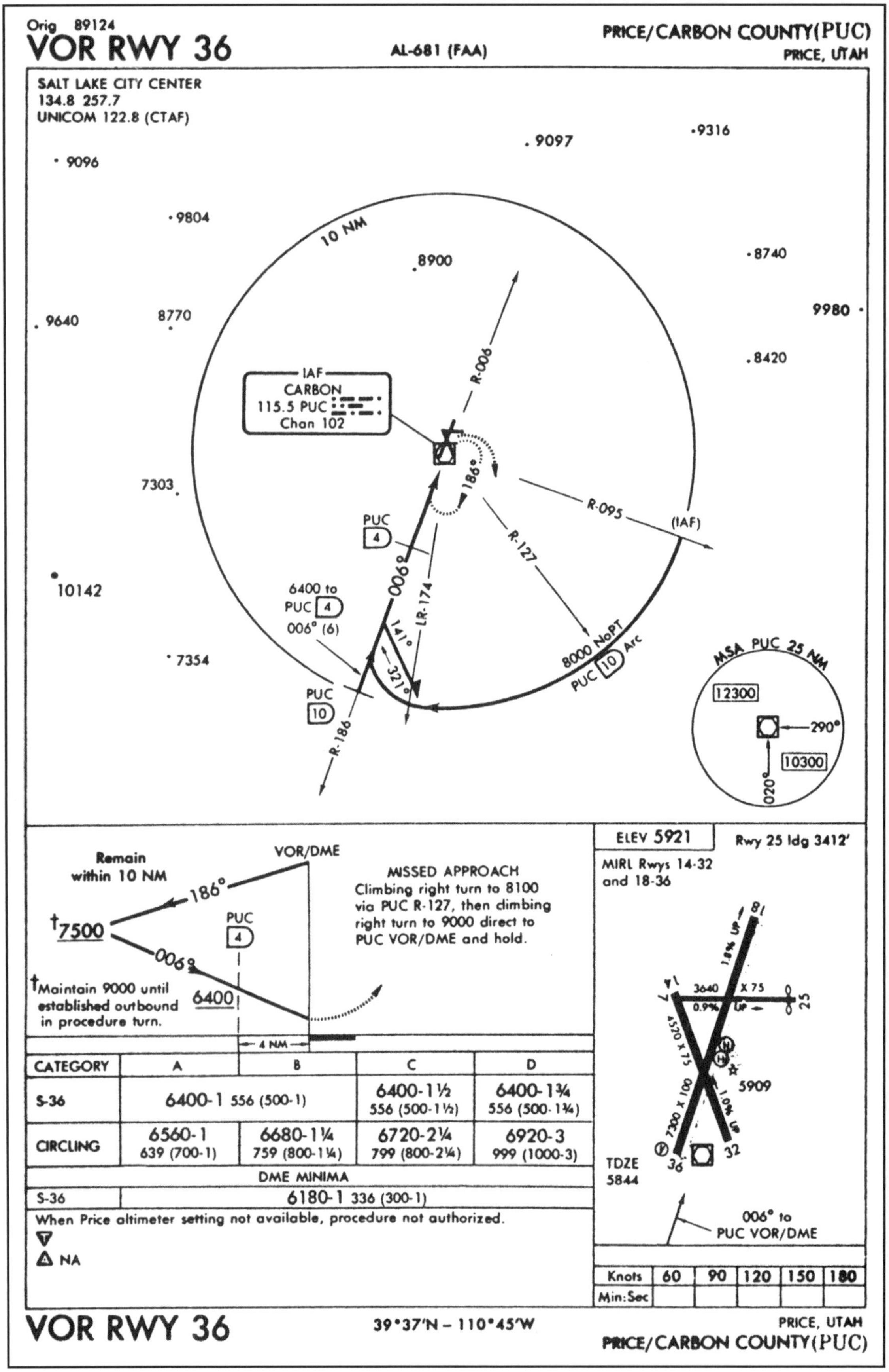

CATEGORY	A	B	C	D
S-36	6400-1 556 (500-1)		6400-1½ 556 (500-1½)	6400-1¾ 556 (500-1¾)
CIRCLING	6560-1 639 (700-1)	6680-1¼ 759 (800-1¼)	6720-2¼ 799 (800-2¼)	6920-3 999 (1000-3)
DME MINIMA				
S-36	6180-1 336 (300-1)			

When Price altimeter setting not available, procedure not authorized.

▽ A NA

Volume 3 – Question 10.2

Amdt 3 91038

LOC RWY 35

AL-5140 (FAA)

DUNCAN/HALLIBURTON FIELD (DUC)
DUNCAN, OKLAHOMA

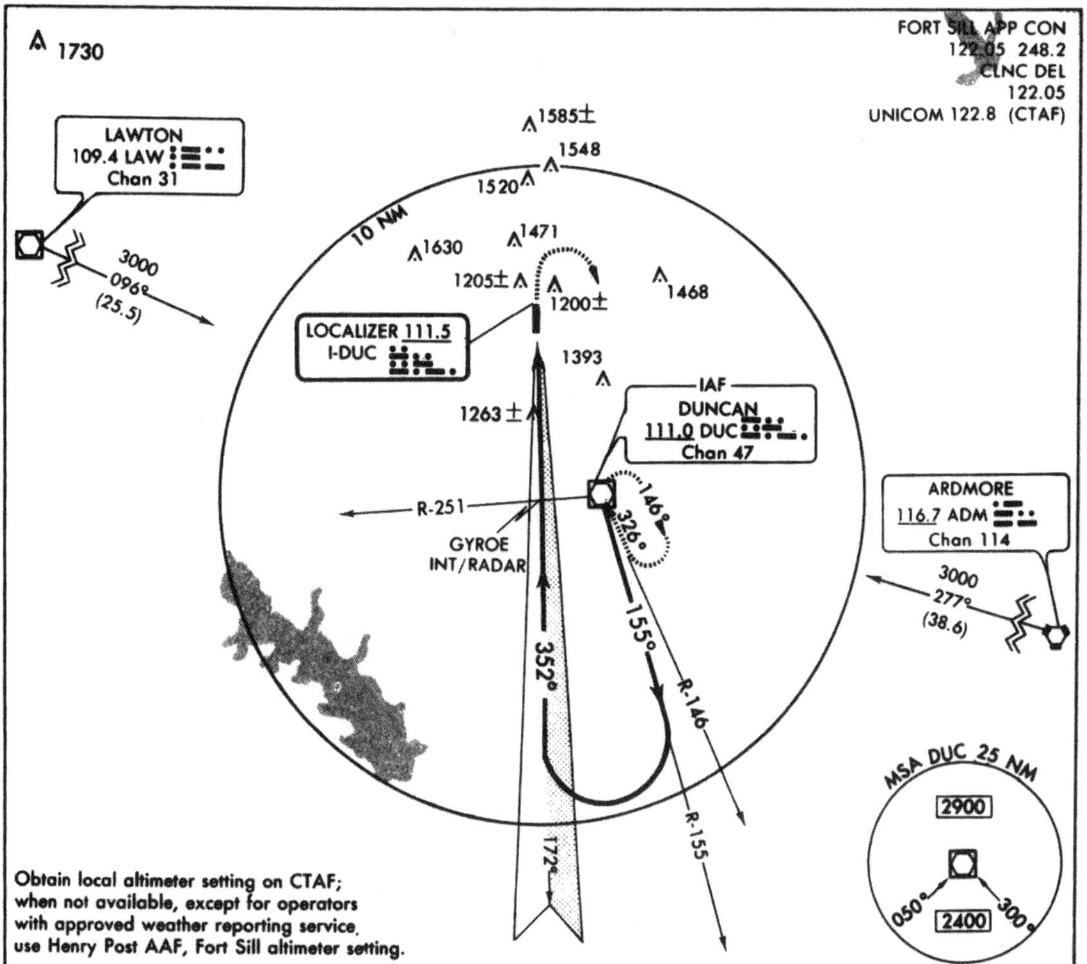

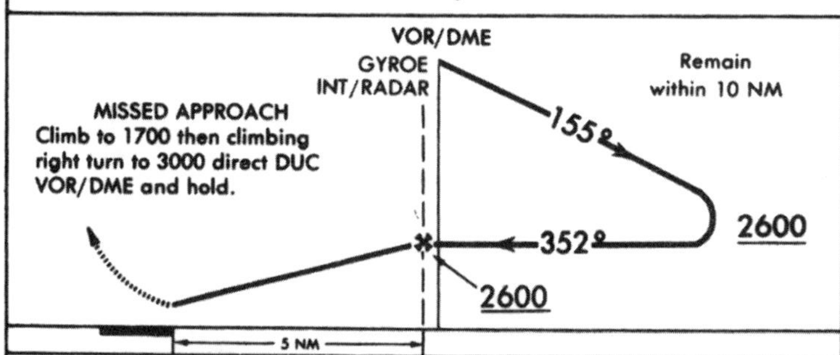

CATEGORY	A	B	C	D
S-35	1520-1 423 (500-1)		1520-1¼ 423 (500-1¼)	
CIRCLING	1560-1 447 (500-1)	1580-1 467 (500-1)	1580-1½ 467 (500-1½)	1780-2 667 (700-2)
HENRY POST AAF, FT. SILL ALTIMETER MINIMUMS				
S-35	1620-1 523 (600-1)		1620-1½ 523 (600-1½)	1620-1¾ 523 (600-1¾)
CIRCLING	1660-1 547 (600-1)	1680-1 567 (600-1)	1680-1½ 567 (600-1½)	1880-2½ 767 (800-2½)

▼
▲ NA

ELEV 1113

MIRL Rwy 17-35
REIL Rwys 17 and 35

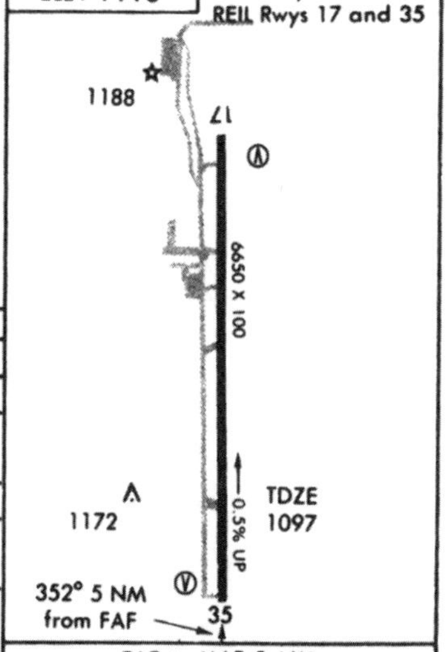

FAF to MAP 5 NM

Knots	60	90	120	150	180
Min:Sec	5:00	3:20	2:30	2:00	1:40

LOC RWY 35

34°28'N-97°58'W

DUNCAN, OKLAHOMA
DUNCAN/HALLIBURTON FIELD (DUC)

Volume 3 – Question 10.3

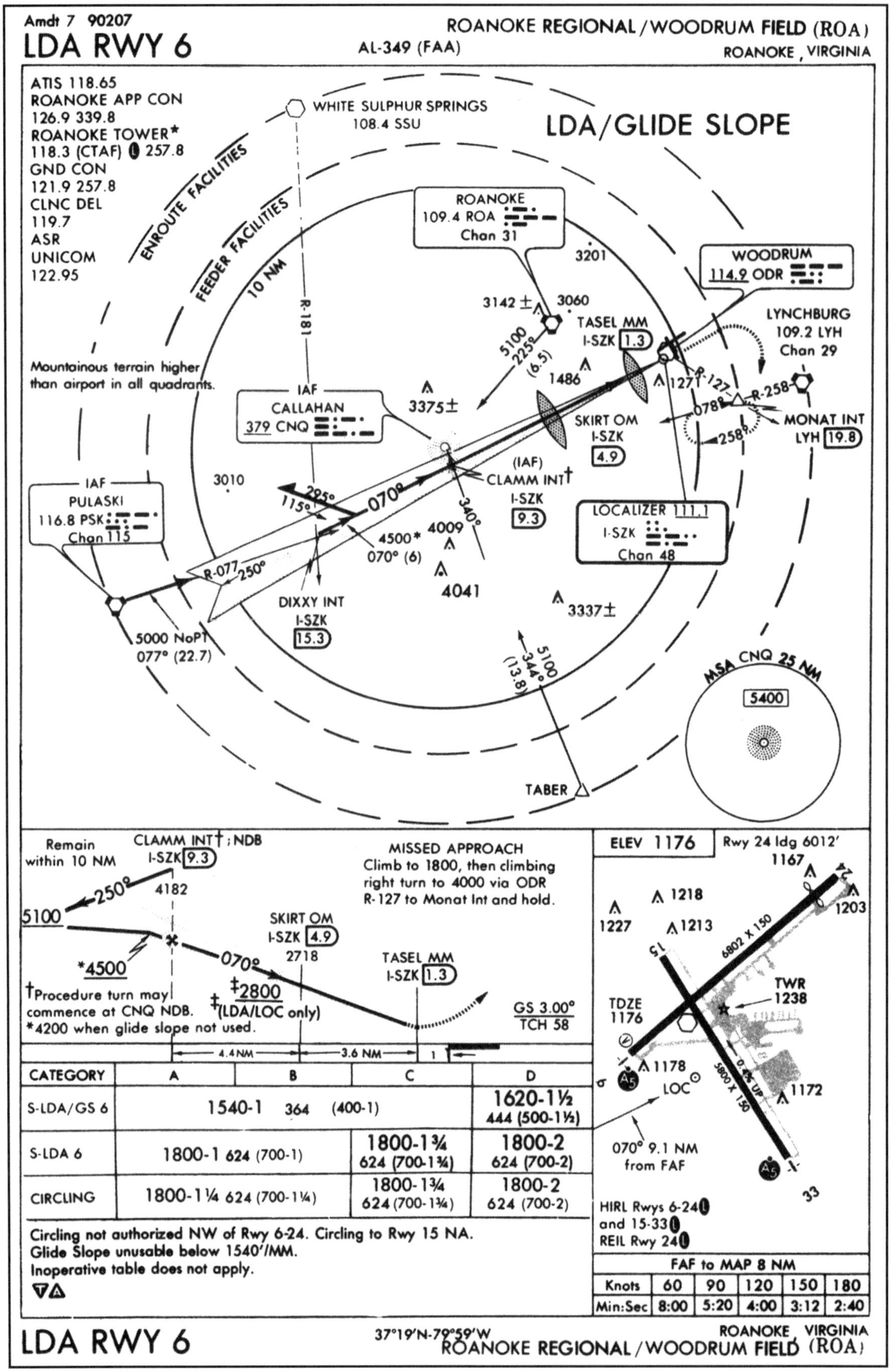

CATEGORY	A	B	C	D
S-LDA/GS 6	1540-1 364 (400-1)			1620-1½ 444 (500-1½)
S-LDA 6	1800-1 624 (700-1)		1800-1¾ 624 (700-1¾)	1800-2 624 (700-2)
CIRCLING	1800-1¼ 624 (700-1¼)		1800-1¾ 624 (700-1¾)	1800-2 624 (700-2)

Circling not authorized NW of Rwy 6-24. Circling to Rwy 15 NA.
Glide Slope unusable below 1540'/MM.
Inoperative table does not apply.

FAF to MAP 8 NM					
Knots	60	90	120	150	180
Min:Sec	8:00	5:20	4:00	3:12	2:40

Volume 3 – Question 10.4

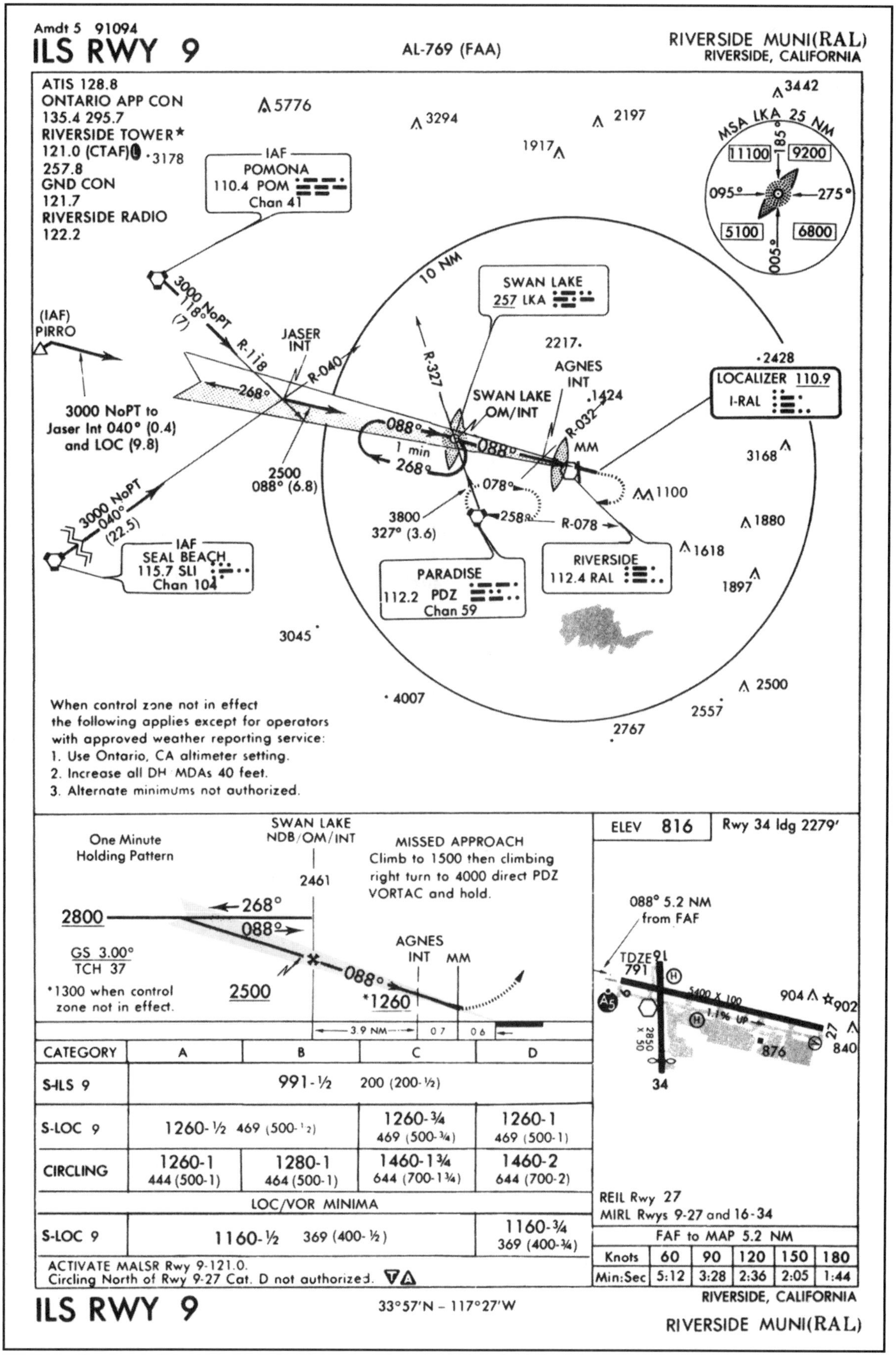

CATEGORY	A	B	C	D
S-ILS 9	991-½ 200 (200-½)			
S-LOC 9	1260-½ 469 (500-½)		1260-¾ 469 (500-¾)	1260-1 469 (500-1)
CIRCLING	1260-1 444 (500-1)	1280-1 464 (500-1)	1460-1¾ 644 (700-1¾)	1460-2 644 (700-2)
LOC/VOR MINIMA				
S-LOC 9	1160-½ 369 (400-½)			1160-¾ 369 (400-¾)

FAF to MAP 5.2 NM					
Knots	60	90	120	150	180
Min:Sec	5:12	3:28	2:36	2:05	1:44

Volume 3 – Question 10.5

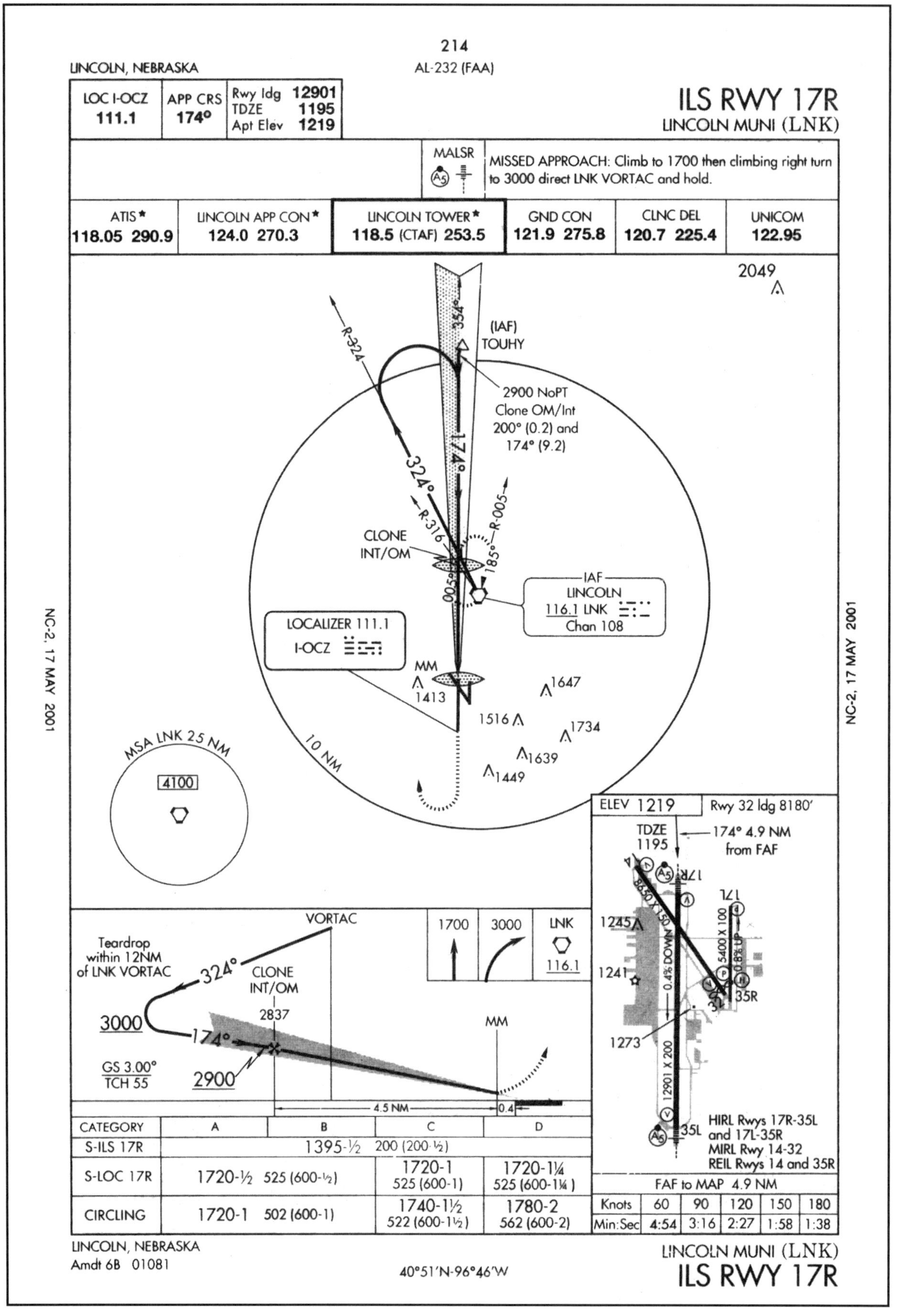

Volume 3 – Question 13.1

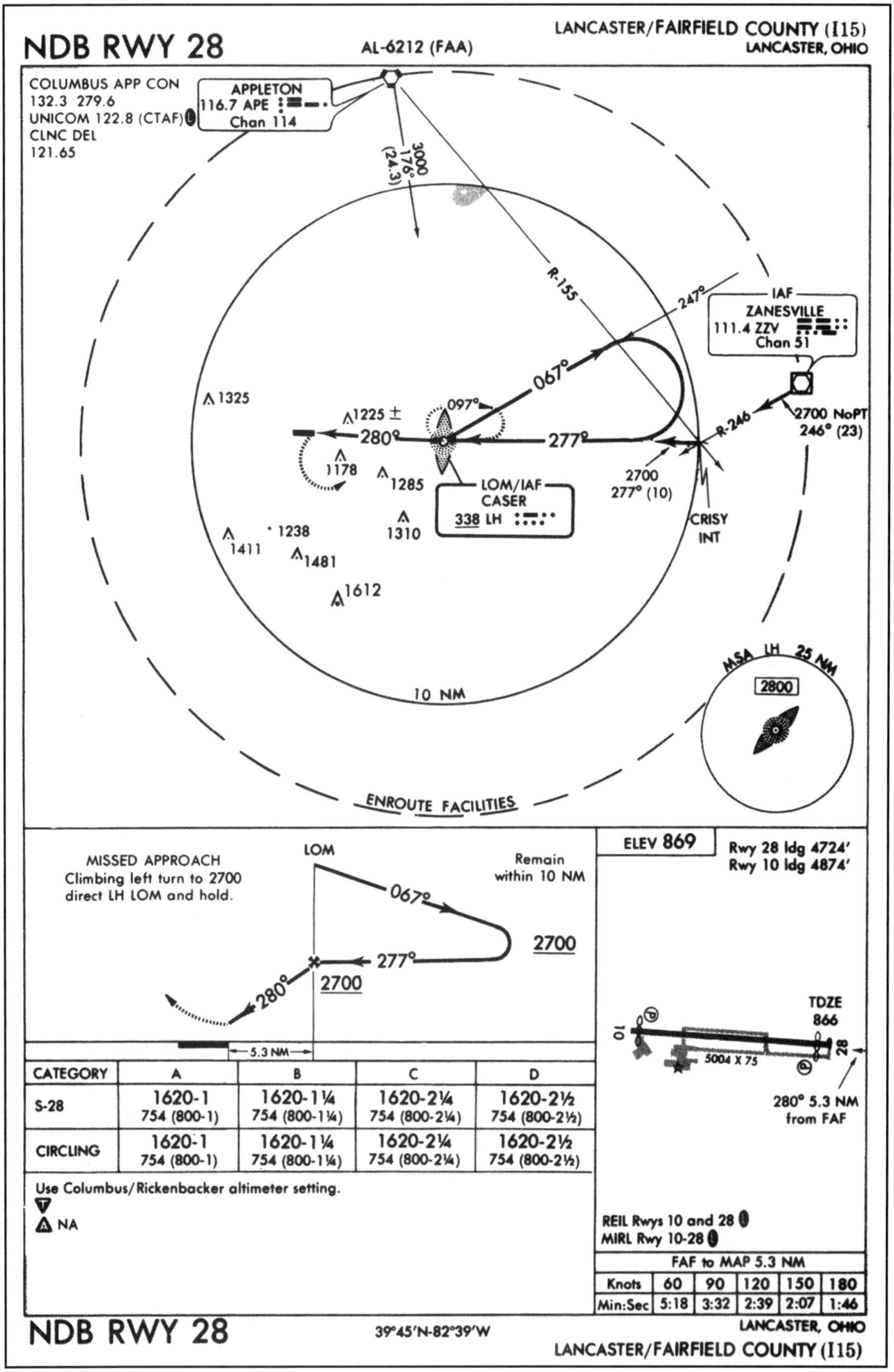

Volume 3 – Question 15.2

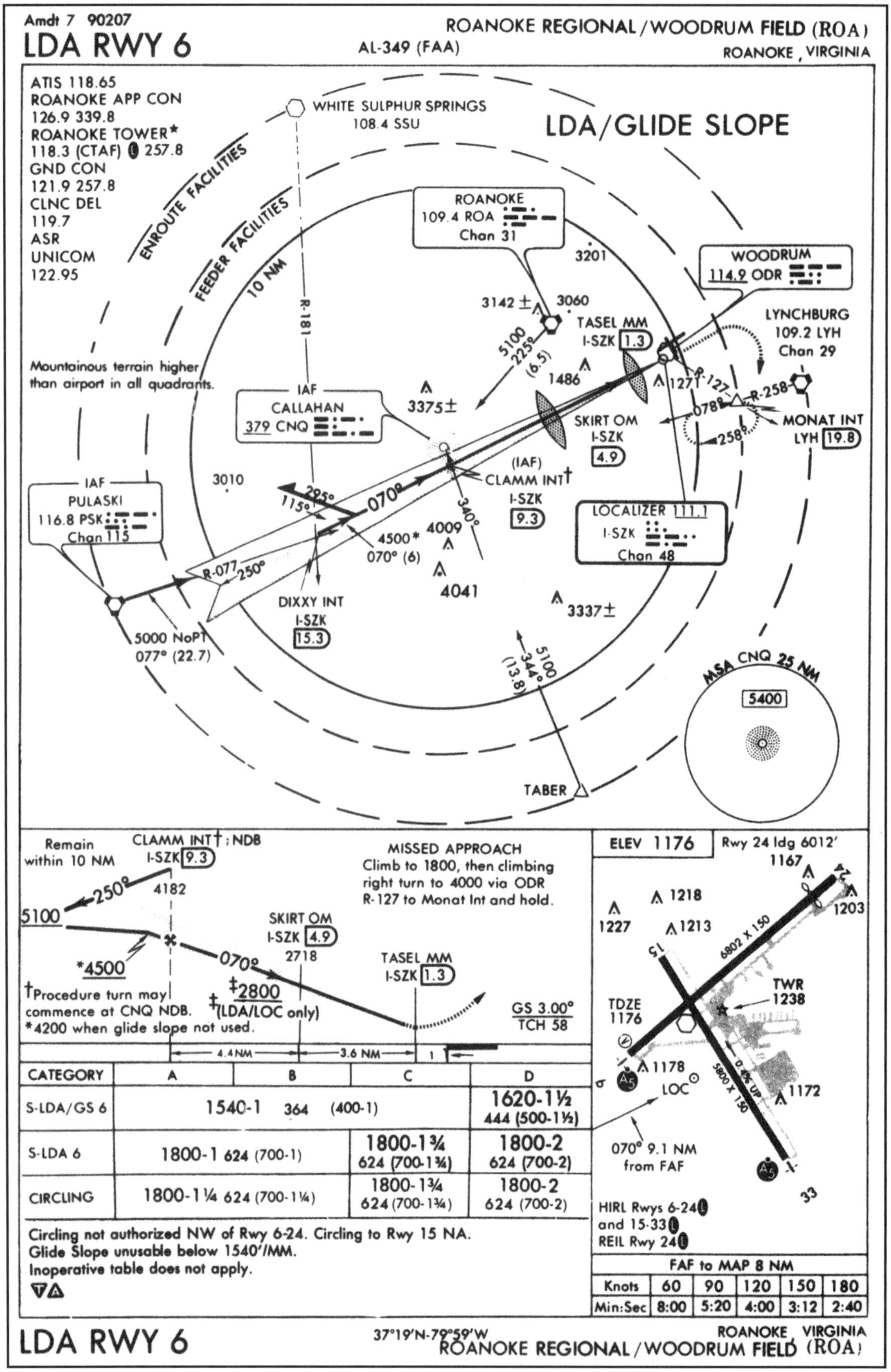

Volume 3 – Question 16.4

Amdt 7 89264

ILS RWY 31

AL-123 (FAA)

DOTHAN (DHN)
DOTHAN, ALABAMA

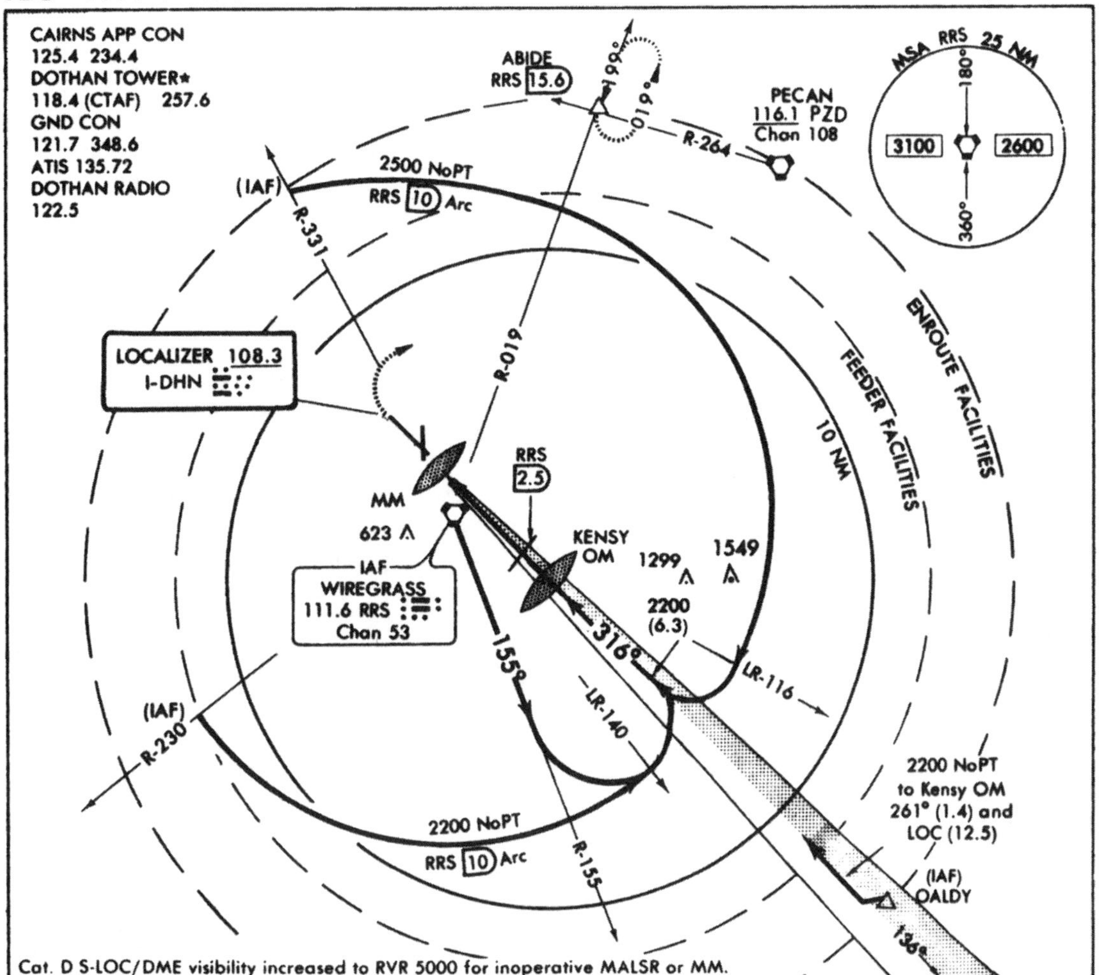

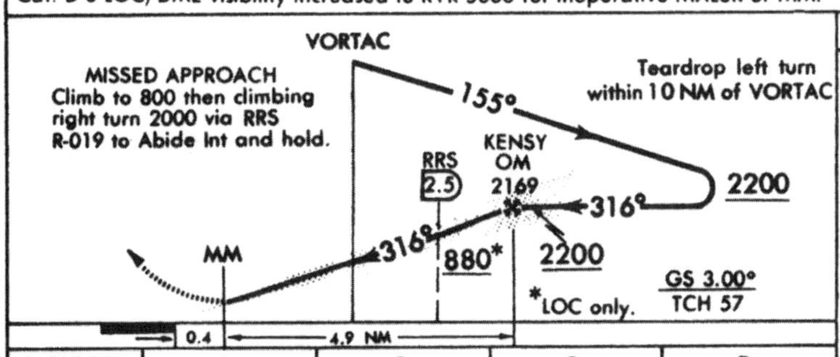

ELEV 401 | HIRL Rwys 13-31 and 18-36

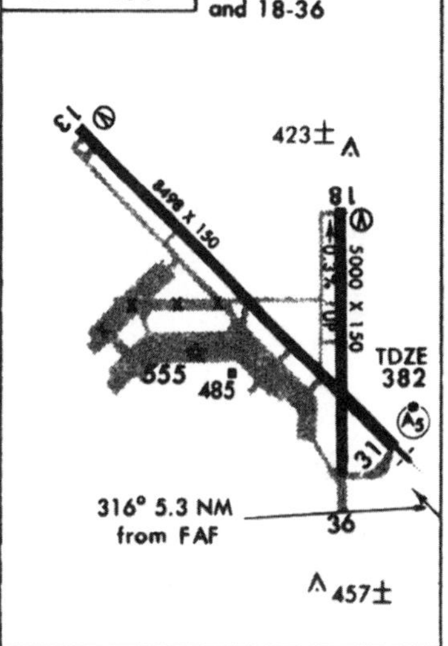

CATEGORY	A	B	C	D
S-ILS 31	582/24 200 (200-½)			
S-LOC 31	880/24 498 (500-½)		880/40 498 (500-¾)	880/50 498 (500-1)
CIRCLING	880-1 479 (500-1)		880-1½ 479 (500-1½)	960-2 559 (600-2)
DME MINIMUMS				
S-LOC 31	720/24 338 (400-½)			720/40 338 (400-¾)
CIRCLING	860-1 459 (500-1)		860-1½ 459 (500-1½)	960-2 559 (600-2)

FAF to MAP 5.3 NM

Knots	60	90	120	150	180
Min:Sec	5:18	3:32	2:39	2:07	1:46

ILS RWY 31

31°19'N – 85°27'W

DOTHAN, ALABAMA
DOTHAN (DHN)

Volume 3 – Question 18.1

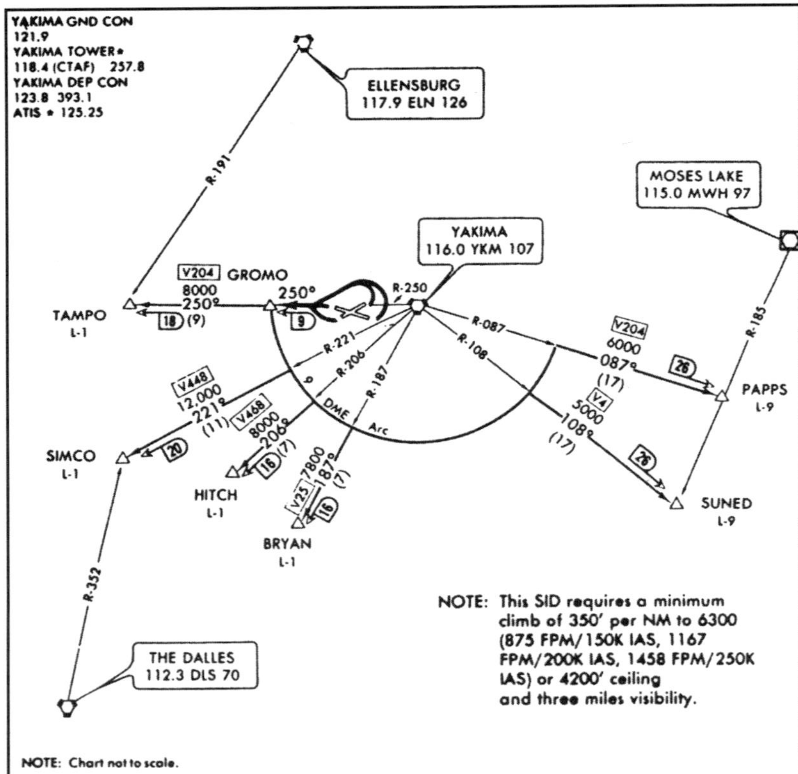

DEPARTURE ROUTE DESCRIPTION

TAKE-OFF RUNWAY 22: Turn right thence . . . .
TAKE-OFF RUNWAY 4, 9: Turn left thence . . . .
TAKE-OFF RUNWAY 27: Climb runway heading thence . . . .

. . . .intercept and proceed via YKM R-250 to the GROMO DME FIX (YKM R-250/9), thence via (assigned transition).

TAMPO TRANSITION (GROMO2.TAMPO): From GROMO DME Fix via V204 to TAMPO INT.
NOTE: Climb in holding pattern east of TAMPO INT, right turns, 250° inbound to 8000' before proceeding on course.

(Continued on next page)

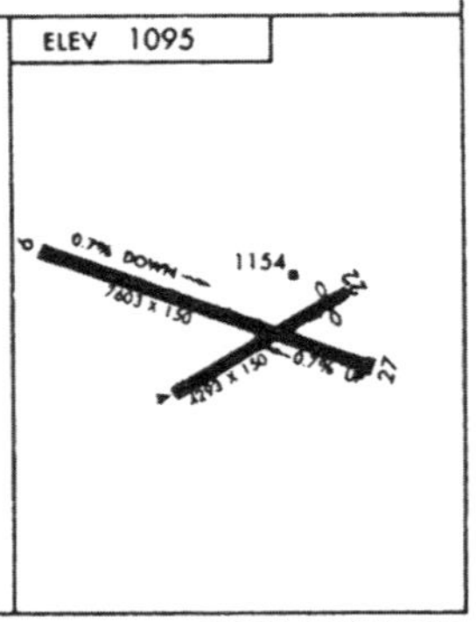

GROMO TWO DEPARTURE (PILOT NAV)
(GROMO2.GROMO)
YAKIMA, WASHINGTON
YAKIMA AIR TERMINAL

(GROMO2.GROMO)
GROMO TWO DEPARTURE (PILOT NAV)
YAKIMA AIR TERMINAL
YAKIMA, WASHINGTON

DEPARTURE ROUTE DESCRIPTION
(Continued)

SIMCO TRANSITION (GROMO2.SIMCO): From over GROMO DME Fix via 9 DME Arc and V448 to SIMCO INT. NOTE: Climb in holding pattern NE of SIMCO INT, right turns, 221° inbound to 12000 feet before proceeding on course.
HITCH TRANSITION (GROMO2.HITCH): From over GROMO DME Fix via 9 DME Arc and V468 to HITCH DME Fix. NOTE: Climb in holding pattern NE of HITCH DME Fix, right turns, 206° inbound to 8000 feet before proceeding on course.
BRYAN TRANSITION (GROMO2.BRYAN): From over GROMO DME Fix via 9 DME Arc and V25 to BRYAN DME Fix.
SUNED TRANSITION (GROMO2.SUNED): From over GROMO DME Fix via 9 DME Arc and V4 to SUNED INT.
PAPPS TRANSITION (GROMO2.PAPPS): From over GROMO DME Fix via 9 DME Arc and V204 to PAPPS INT.

GROMO TWO DEPARTURE (PILOT NAV)
(GROMO2.GROMO)
YAKIMA, WASHINGTON
YAKIMA AIR TERMINAL

§ **YAKIMA AIR TERMINAL** (YKM) 3 S GMT–8(–7DT) 46°34'06"N 120°32'33"W — **SEATTLE** H-1A, L-1C, 9A **IAP**
1095 B S4 **FUEL** 80, 100, 100LL, JET A OX 1, 3 CFR Index A
RWY 09-27: H7603X150 (ASPH-PFC) S-95, D-110, DT-175 HIRL .6% up NW
RWY 09: VASI(V4L)—GA 3.0° TCH 50'. Antenna. **RWY 27:** MALSR. Trees.
RWY 04-22: H4293X150 (ASPH) S-70, D-80, DT-120 MIRL
RWY 04: Trees. **RWY 22:** VASI(V2L)—GA 3.03° TCH 61'. Thld dsplcd 469'. Tree.
AIRPORT REMARKS: Attended continuously. CLOSED 0600-1400Z‡ to Part 121 non-scheduled operations and Part 121 scheduled charters operations without prior approval, call (509) 575-6150/6014. Bird hazard. ACTIVATE MALSR Rwy 27—118.4.
COMMUNICATIONS: CTAF 118.4 ATIS 125.25 (1400-0600Z‡) **UNICOM** 122.95
SEATTLE FSS (SEA) LC 965-1840 NOTAM FILE YKM
RCO 122.5 122.1R 116.0T (SEATTLE FSS)
APP/DEP CON 123.8 (1400-0600Z‡) **SEATTLE CENTER APP/DEP CON** 132.6 (0600-1400Z‡)
TOWER 118.4 (1400-0600Z‡) **GND CON** 121.9
VFR ADVSY SVC 123.8
RADIO AIDS TO NAVIGATION:
(H) **VORTAC** 116.0 YKM Chan 107 46°34'14"N 120°26'36"W 244° 3.6 NM to fld. 980/21E
DME portion unusable:

350°-065° beyond 20 NM below 9000'	025°-065° beyond 15 NM below 8000'
115°-135° beyond 25 NM below 7000'	135°-150° beyond 30 NM below 7500'
195°-230° beyond 22 NM below 8000'	300°-325° beyond 29 NM below 9000'
325°-335° beyond 20 NM below 9000'	335°-350° beyond 25 NM below 10,000'

VOR portion unusable:

350°-080° beyond 25 NM below 9000'	025°-035° beyond 5 NM below 10,000'
080°-105° beyond 35 NM below 6000'	105°-107° beyond 25 NM below 6000'
109°-135° beyond 25 NM below 6000'	135°-180° beyond 30 NM below 7500'
195°-225° beyond 30 NM below 8500'	305°-335° beyond 30 NM below 9000'

DONNY NDB (LOM) 371 YK 46°31'33"N 120°22'16"W 269° 7 NM to fld
ILS 110.1 I-YKM Rwy 27 LOM DONNY NDB. ILS unmonitored when tower closed. Localizer unusable inside threshold.

Volume 3 – Question 18.1

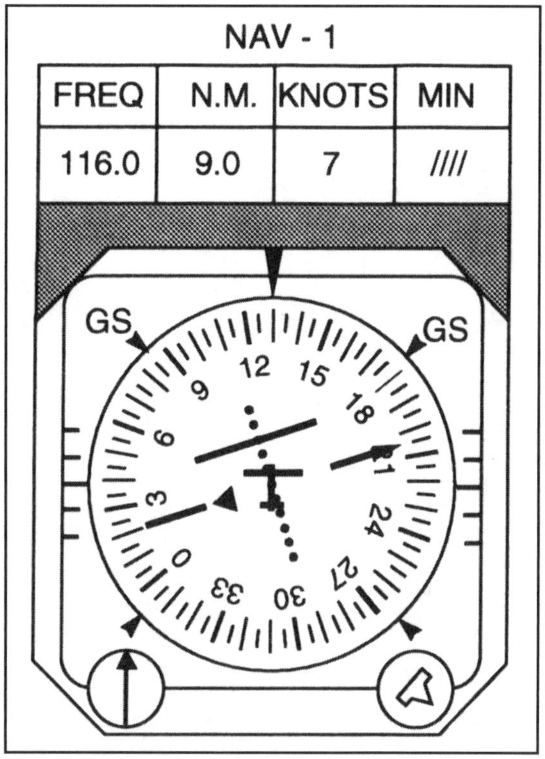

Volume 3 – Question 18.2

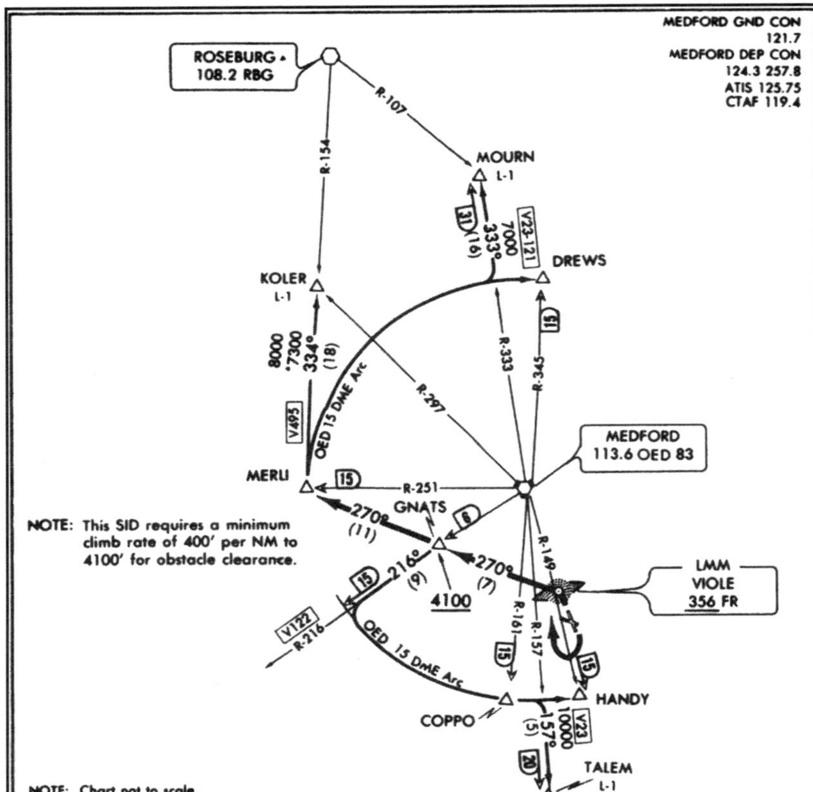

▼ DEPARTURE ROUTE DESCRIPTION

Climb direct to the VIOLE ILS Middle Compass Locator (south take-off turn right), then climb on the 270° magnetic bearing from the LMM to GNATS INT, cross GNATS INT at or above 4100; thence via (transition) or (route).

COPPO TRANSITION (GNATS1.COPPO): Turn left via R-216 to 15 DME Fix, thence turn left via MEDFORD 15 DME Arc to COPPO DME Fix.

KOLER TRANSITION (GNATS1.KOLER): Continue via 270° magnetic bearing from the LMM to MERLI INT, turn right via ROSEBURG R-154 to KOLER INT.

(Continued on next page)

ELEV 1331

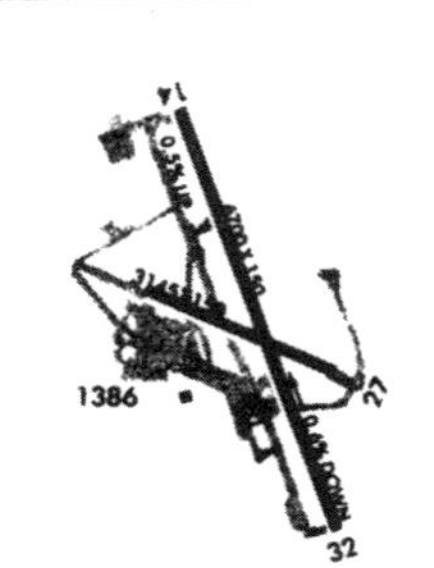

GNATS ONE DEPARTURE (GNATS1.GNATS) MEDFORD, OREGON MEDFORD-JACKSON CO

GNATS ONE DEPARTURE (GNATS1.GNATS)

MEDFORD-JACKSON CO
MEDFORD, OREGON

DEPARTURE ROUTE DESCRIPTION (Continued)

MOURN TRANSITION (GNATS1.MOURN): Continue via 270° magnetic bearing from the LMM to MERLI INT, turn right via MEDFORD 15 DME ARC to intercept V23-121 to MOURN INT.

DREWS TRANSITION (GNATS1.DREWS): Continue via 270° magnetic bearing from the LMM to MERLI INT, turn right via MEDFORD 15 DME ARC to DREWS INT.

TALEM TRANSITION (GNATS1.TALEM): Turn left via MEDFORD R-216 to 15 DME Fix thence turn left via MEDFORD 15 DME ARC to Intercept V23 to TALEM INT.

HANDY TRANSITION (GNATS1.HANDY): Turn left via MEDFORD R-216 to 15 DME Fix, thence turn left via MEDFORD 15 DME Arc to HANDY DME Fix.

GNATS ONE DEPARTURE (GNATS1.GNATS) MEDFORD, OREGON MEDFORD-JACKSON CO

§ **MEDFORD-JACKSON CO** (MFR) 3 N GMT–8(–7DT) 42°22′21″N 122°52′17″W **KLAMATH FALLS**
1331 B S4 **FUEL** 80, 100, 100LL, JET A1 + OX 1, 3 CFR Index B **H-1A, L-1A**
RWY 14-32: H6700X150 (ASPH-PFC) S-200, D-200, DT-400 HIRL .5% up S **IAP**
RWY 14: MALSR. Trees. **RWY 32:** REIL. VASI(V4L)—GA 3.0°TCH 49′. Road.
RWY 09-27: H3145X150 (ASPH) S-50, D-70, DT-108 MIRL
RWY 27: Road.
AIRPORT REMARKS: Attended continuously. CLOSED to unscheduled Part 121 air carriers operation, without prior approval, call 503-776-7222. Night refueling delay sunset-1500Z‡, ctc TOWER. Rwy 09-27 clsd to acft over 12,500 lbs GWT. Rwy 09/27 CLOSED when tower clsd. Rwy lgts 14/32 operate med ints when tower closed. ACTIVATE MALSR 14—119.4. Flocks of large waterfowl in vicinity Nov-May
COMMUNICATIONS: CTAF 119.4 **ATIS** 125.75 **UNICOM** 122.95
NORTH BEND FSS (OTH) LC 773-3256. NOTAM FILE MFR.
RCO 122.65 122.1R 113.6T (NORTH BEND FSS)
APP CON 124.3 (1400-0800Z‡) **DEP CON** 124.3 (1400-0800Z‡)
SEATTLE CENTER APP/DEP CON 125.3 (0800-1400Z‡)
TOWER 119.4 (1400-0800Z‡) **GND CON** 121.7
VFR ADVSY SVC ctc TOWER
RADIO AIDS TO NAVIGATION: NOTAM FILE OTH. VHF/DF ctc Medford TOWER
(H) ABVORTAC 113.6 ■ OED Chan 83 42°28′47″N 122°54′43″W 146°6.1 NM to fld. 2080/19E
VORTAC unusable:
160°-165° beyond 35 NM below 8900′ 280°-345° beyond 30 NM below 6500′
198°-205° beyond 35 NM below 8500′ 345°-360° beyond 35 NM below 6800′
250°-280° beyond 25 NM below 6100′
PUMIE NDB (LOM) 373 MF 42°27′04″N 122°54′44″W 140°4.5 NM to fld. NOTAM FILE MFR
LOM unusable 150°-165° and 260°-265° beyond 5 miles.
VIOLE NDB (LMM) 356 FR 42°23′22″N 122°52′47″W 140°0.5 NM to fld. NOTAM FILE MFR
LMM unusable 305°-335° beyond 10 NM. all altitudes
ILS/DME 110.3 I-MFR Chan 40 Rwy 14 LOM PUMIE NDB. LMM VIOLE NDB. ILS unmonitored when tower closed. Localizer unusable inside threshold.

Volume 4 - En Route IFR

Charts and Airways

1) Low Altitude En Route Charts (L Charts) contain information for IFR flights on Victor Airways. Generally, Victor Airways are numbered even when they run east-west and odd when they run north-south. IFR flights that are not flown on an airway are called direct routes. Victor Airways extend up to but do not include 18,000 ft. MSL.

2) High Altitude En Route Charts (H Charts) contain information on High Altitude Airways, or Jet Routes. High Altitude Airways begin at 18,000 ft. MSL and continue up to FL450. Airplanes flying above FL450 fly point to point, called direct navigation.

3) Area Charts are printed for areas congested with airways and navigation information. Areas that have published area charts are denoted on the L charts. L charts depict all the necessary information for flight through the area, while Area Charts contain the information necessary for arrivals and departures inside the area.

Airway Limits and Minimums

1) The MOCA, or Minimum Obstruction Clearance Altitude is at least 1000 ft. above the highest obstruction within 4 N.M. of the airway centerline. In mountainous terrain, the clearance is increased to 2000 ft, although under certain circumstances, it can be dropped to 1500 ft. It also allows for a satisfactory navigational signal when within 22 N.M. of a station. On direct flights, pilots are responsible for obstacle clearance. Fixes named on direct flights are compulsory reporting points.

2) The MEA, or Minimum En Route Altitude, allows for obstacle clearance and satisfactory navigational signal reception throughout the airway, and for the identification of fixes on the airway.
   a) Some fixes cannot be identified at the MEA. This is because the signal from the side station cannot be received at the MEA. An MRA, or Minimum Reception Altitude, is published for these fixes.
   b) An MEA Gap denotes a break in the navigation signal coverage, or a signal, which does not meet airway standards. When flying through an MEA gap, the proper heading should be maintained until reception resumes.
   c) An MCA, or Minimum Crossing Altitude is the lowest altitude at which an aircraft may cross a fix when transitioning to a higher MEA.

3) MAAs, or Maximum Authorized Altitudes, are applied to some Victor Airways in order to keep aircraft out of conflicting airspace, such as overhanging layers of Class B airspace.

4) COPs, or Changeover Points, are points along the airway between two adjacent navigational facilities or waypoints where a switch in navigational aid should occur. Midpoint COPs are not charted, COPs located other than at midpoints are denoted on the chart. The COP on a course that changes direction is located at the bend in the airway.

5) Preferred Routes are established between busier airports, and are set up to increase efficiency and capacity. They normally extend through more than one ARTCC, and are recommended, not required. Preferred Routes are listed in the Airport/Facility Directory.

Instrument Departure Procedures (DPs)

1) An Instrument Departure Procedure is an ATC coded departure procedure that has been established at certain airports to simplify clearance delivery procedures. DPs are published by the FAA in the Terminal Procedures Publication. There are two basic types of DPs:
   a) Obstacle Departure Procedures (ODPs).
      i) Printed either textually or graphically.
      ii) Provide obstruction clearance via the least onerous route from the terminal area to the en route structure.
      iii) May be flown without ATC clearance unless an alternate DP has been specifically assigned by ATC.
   b) Standard Instrument Departures (SIDs).
      i) Always printed graphically.
      ii) ATC procedures printed for pilot/controller use in graphic form to provide obstruction clearance and a transition from the terminal area to the en route structure.
      iii) Primarily designed for system enhancement and to reduce pilot/controller workload.
      iv) ATC clearance must be received prior to flying a SID.

Standard Terminal Arrivals (STARs)

1) A Standard Terminal Arrival is issued as a transition from a point en route to an approach position, and is used to simplify clearance delivery procedures. STARs are published by the National Aeronautical Charting Office in the Terminal Procedures Publication.

Transition Areas

1) Transition areas are Class E airspace areas extending upward from 700 feet or more AGL when designated in conjunction with an instrument approach procedure, or from 1,200 feet AGL or more when designated in conjunction with airway route structures or segments. They are designated to maintain controlled airspace while transitioning between the terminal and en route environments, and, unless specified otherwise, terminate at the base of the overlying controlled airspace.
    a) The term "Transition Areas" is generally no longer used. These Class E areas are now described by the phrase, "Airspace used for transition." The older term still appears in some FAA information, including the Instrument Knowledge Test, and is included here for completeness.

Alternate Airports

1) **Fuel Requirements** -- 14 CFR §91.167 requires that an aircraft operating in IFR conditions, given weather reports and forecasts, must have enough fuel to:
    a) Complete the flight to the first airport of intended landing, and
    b) Fly from that airport to the alternate airport, and thereafter,
    c) Fly for another 45 minutes at cruising speed
2) **Alternate Airport Minimums** -- For flight planning purposes, 14 CFR §91.169 requires that an airport must have the following standard terminal forecast minimums at the ETA in order to be used as an alternate:
    a) For a precision approach, a 600 foot ceiling and 2 statute miles visibility.
    b) For a nonprecision approach, an 800 foot ceiling and 2 statute miles visibility.
    c) For a non-instrument approach, there must be a forecast for VFR conditions from the MEA until landing.
3) An alternate is not mandatory if the first airport of intended landing has an IAP and has been forecast to have at least a 2000 foot ceiling and 3 miles of visibility for at least one hour before and one hour after the ETA. However, it is not necessary to go to the specified alternate airport if unable to land at the intended destination.

Weather Data

1) **Area Forecasts** cover a fairly large area; six cover the contiguous United States. They are a good source of information for en route conditions. Area Forecasts are issued three times per day, and include a 12-hour forecast, as well as a 6-hour outlook. They are divided into four sections:
    a) Communications and Product Headers
    b) Precautionary Statements
    c) Synopsis
    d) VFR Clouds and Weather
2) **Terminal Aerodrome Forecasts** predict weather conditions expected within 5 S.M. of the terminal. Use of the code "VC" (vicinity) applies to weather conditions expected to occur from between 5 to 10 S.M. from the airport. TAFs are issued four times daily at 0000Z, 0600Z, 1200Z and 1800Z and usually cover a 24-hour period.
3) **Winds Aloft Forecasts** provide an estimate of wind direction and speed, as well as temperature for certain altitudes. The conditions are normally given in a six-digit code, with the first two digits representing tens of degrees of wind direction. The two middle digits represent the wind speed, and the final two digits (separated by a plus or minus sign) represent the temperature. The wind direction is given in reference to true north, wind speed is given in knots, and temperature is given in Celsius.
    a) NOTE: Since wind direction and speed must conform to a four digit code, special codes are used for light and variable winds, as well as wind speeds that are over 99 knots. Light and variable winds are depicted in the four digit code as "9900". For windspeeds over 99 knots, 50 is added to the wind direction digits. For example, for a wind from 230° with a speed of 108 knots will be depicted in the four digit code as "7308".

4) **Aviation Routine Weather Report**, or METAR report, as the name implies, is an actual observation taken from the surface of the airport every hour. If rapid changes occur in the weather, special report observations are taken. Surface Weather Observations will contain any of the following information that is pertinent to the observation:
    a) Type of Report -- METAR or SPECI (special)
    b) Station designator -- ICAO identifier
    c) Time of report -- reported in UTC
    d) Wind information -- Direction in tens of degrees from true north and wind speed in knots
    e) Visibility -- reported in statute miles or Runway Visual Range (RVR) in feet
    f) Weather and obstructions to visibility
    g) Sky condition -- Height of ceiling and other layers, and amount of coverage of layers
    h) Temperature and dew point -- reported in degrees Celsius
    i) Altimeter setting -- given in inches of mercury
    j) Remarks -- any significant data not reported above
        i) NOTE: Pertinent observation information is broadcast over ATIS.
5) **Pilot Reports** (PIREPs) give actual conditions encountered by aircraft in flight. PIREPs usually contain information concerning cloud tops, icing, visibility, and turbulence.
6) **Radar Weather Reports** (SD) observe thunderstorms and general areas of precipitation, indicating its type, intensity, trend, and location. Also included is the echo top of the precipitation and, if significant, the base echo. Radar Summary Charts report the same information, but are presented on a map instead of textual form. Both serve to show lines and areas of precipitation and thunderstorms. Weather radar cannot detect ceilings, fog, or clouds.
7) **AIRMETs** are warnings of weather hazards that, although of possible interest to all aircraft, are of special interest to light aircraft. AIRMETs are issued for moderate icing and turbulence, high surface winds, and low visibility areas.
8) **SIGMETs** are weather advisories that deal with weather significant to the safety of all aircraft. They cover severe and extreme turbulence, severe icing, widespread sandstorms and dust as well as volcanic eruptions. SIGMETs are available from FSS. Convective SIGMETs are issued specifically for tornadoes, severe and/or embedded thunderstorms, lines of thunderstorms, strong thunderstorms with a large area of coverage, and areas of hail 3/4 inch in diameter or greater.
9) **Constant Pressure Analysis Charts** are upper air weather maps, and show areas of constant pressure. They are prepared twice daily, and are prepared for six pressure levels (850, 700, 500, 300, 250 and 200 Mb). Constant pressure charts give information on observed temperature, the temperature/dew point spread, wind, and height of the pressure surface and height changes over the previous 12 hours.
10) **Weather Depiction Charts** are computer prepared from METARs to give a broad overview of observed flying category conditions at the valid time of the chart. This computer-prepared chart is valid at the time of the plotted data, and is prepared at 3-hour intervals.

Volume 4 – Question 2.1

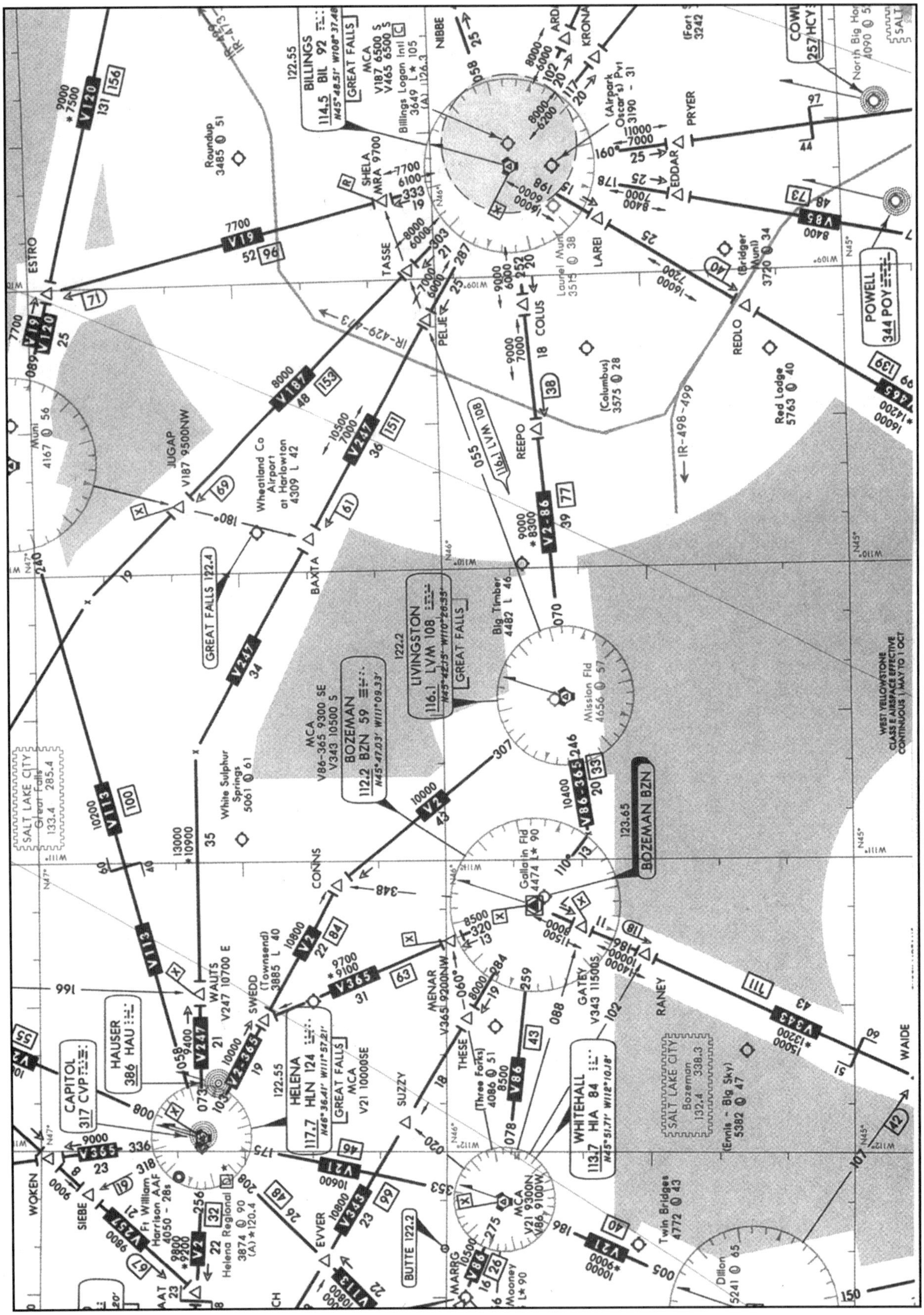

Volume 4 – Question 2.3

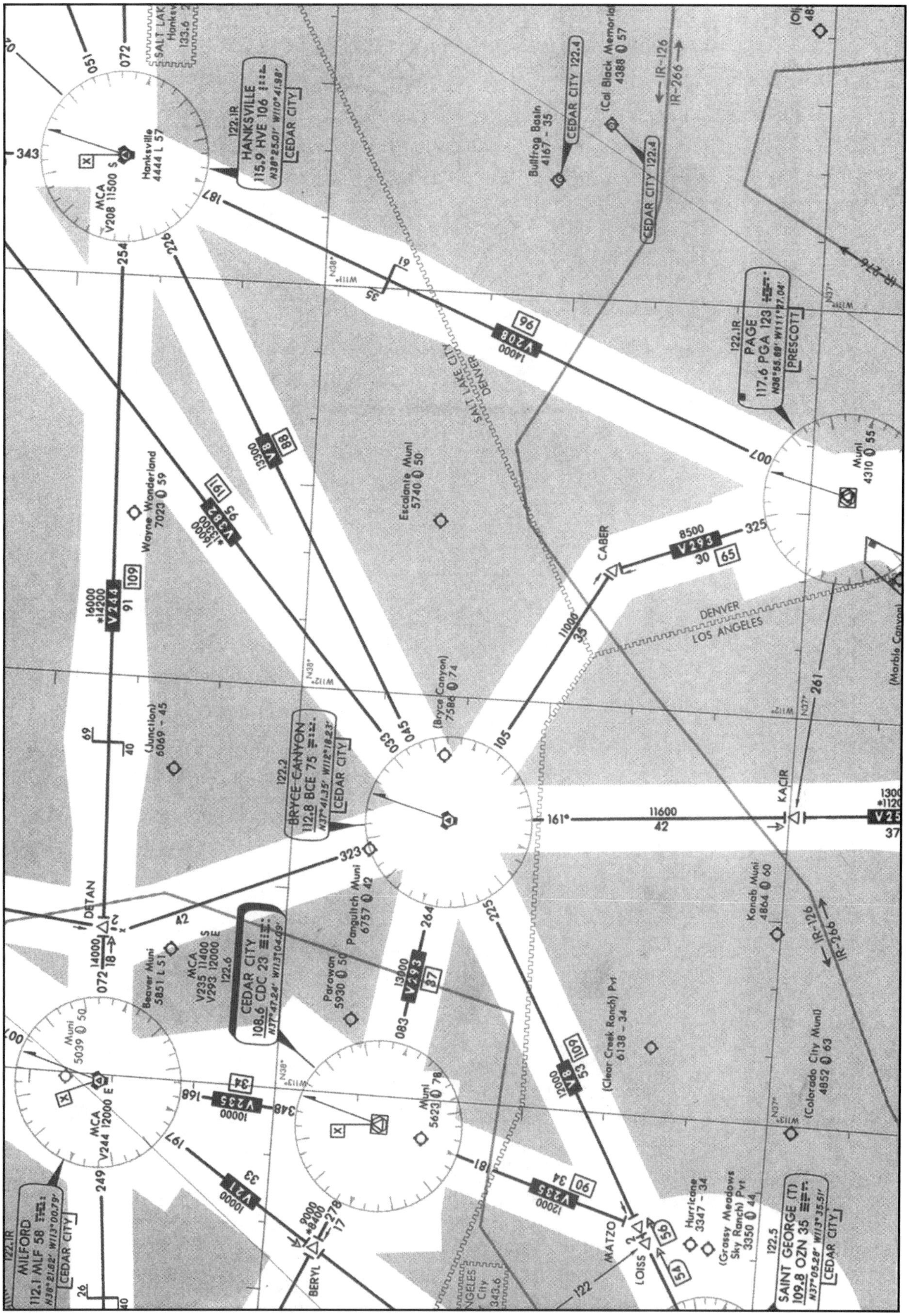

Volume 4 – Question 2.6

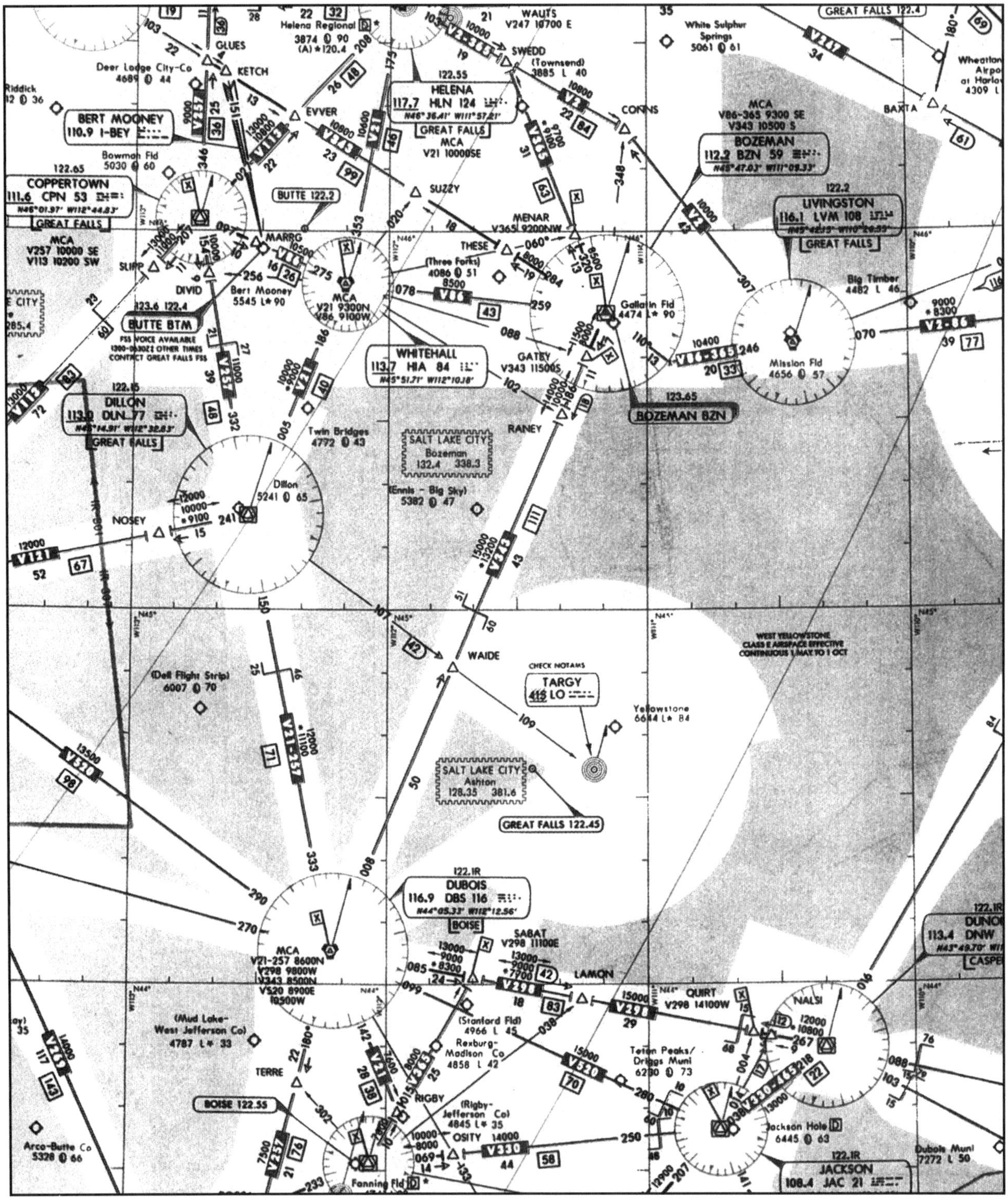

Volume 4 – Question 2.7

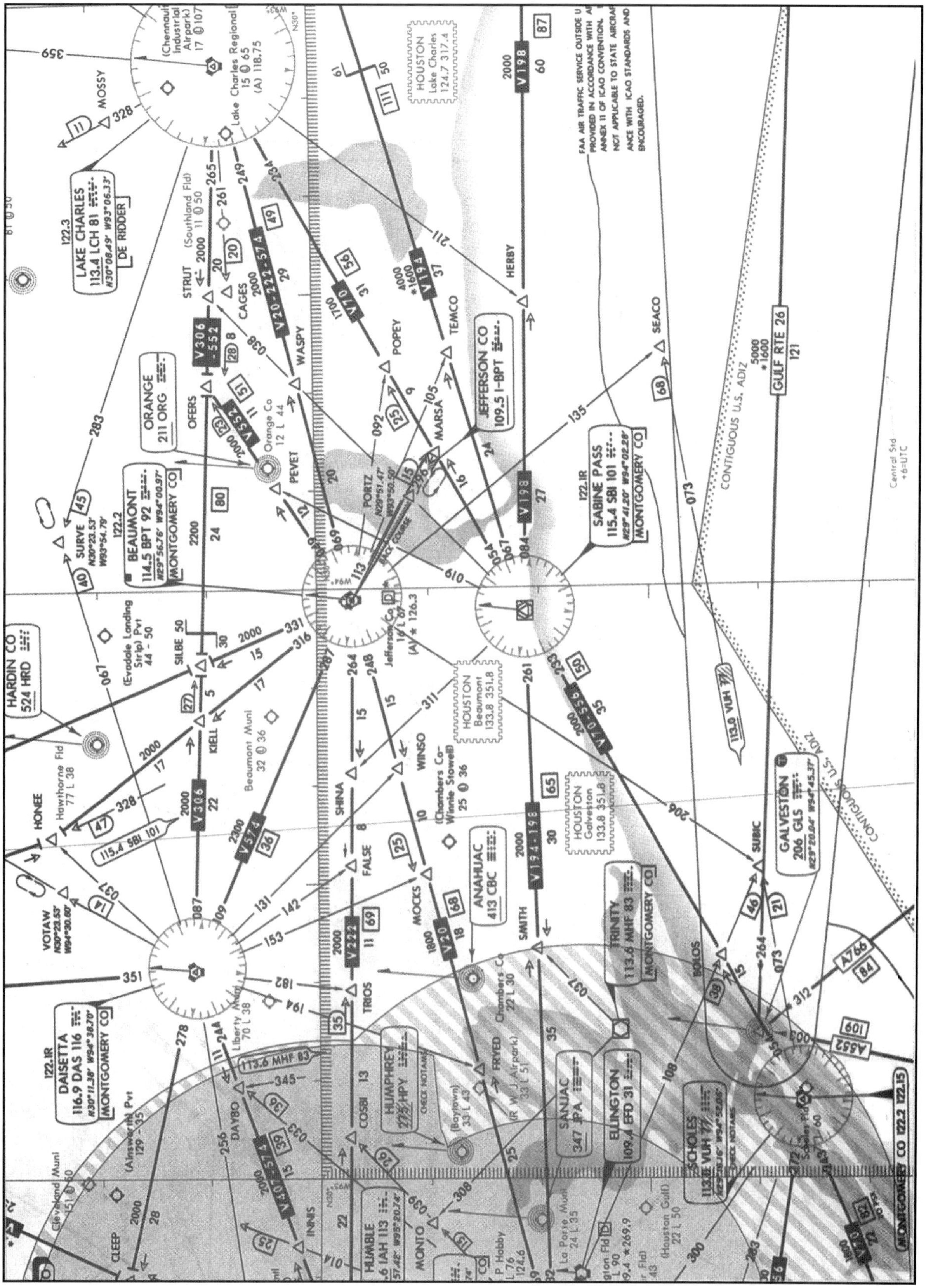

Volume 4 – Questions 2.8 & 3.1

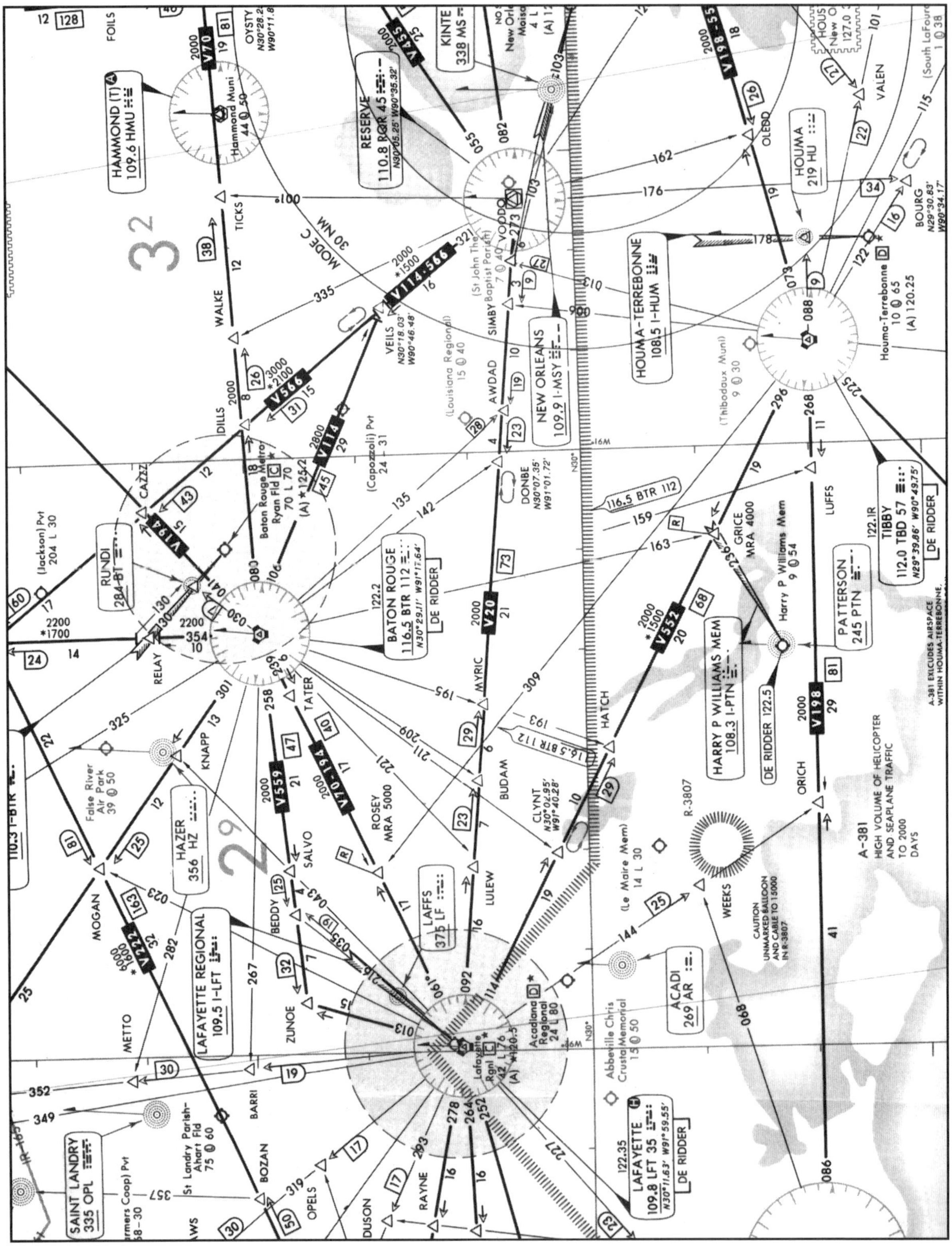

Volume 4 – Question 3.2

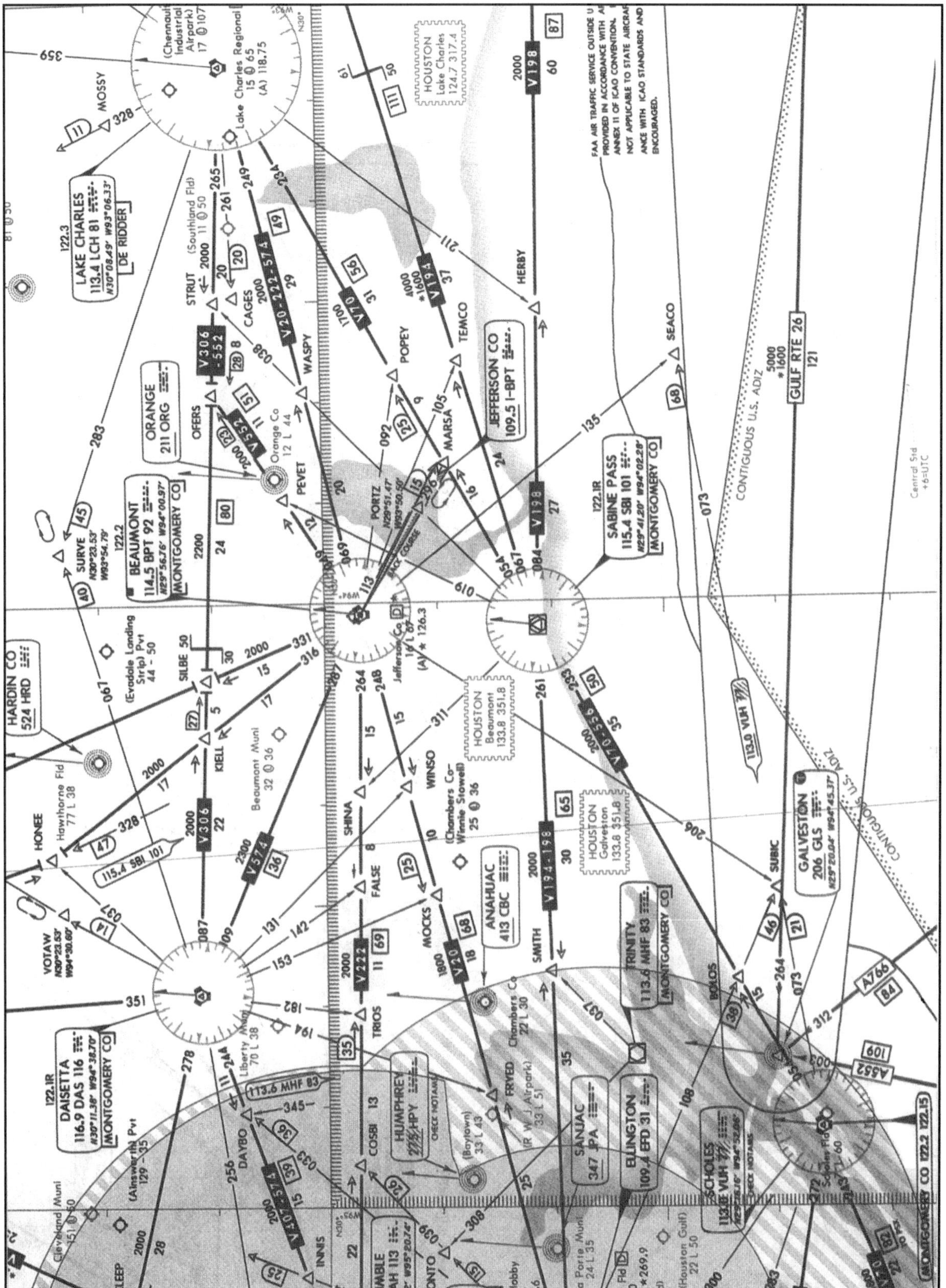

Volume 4 – Question 8.1

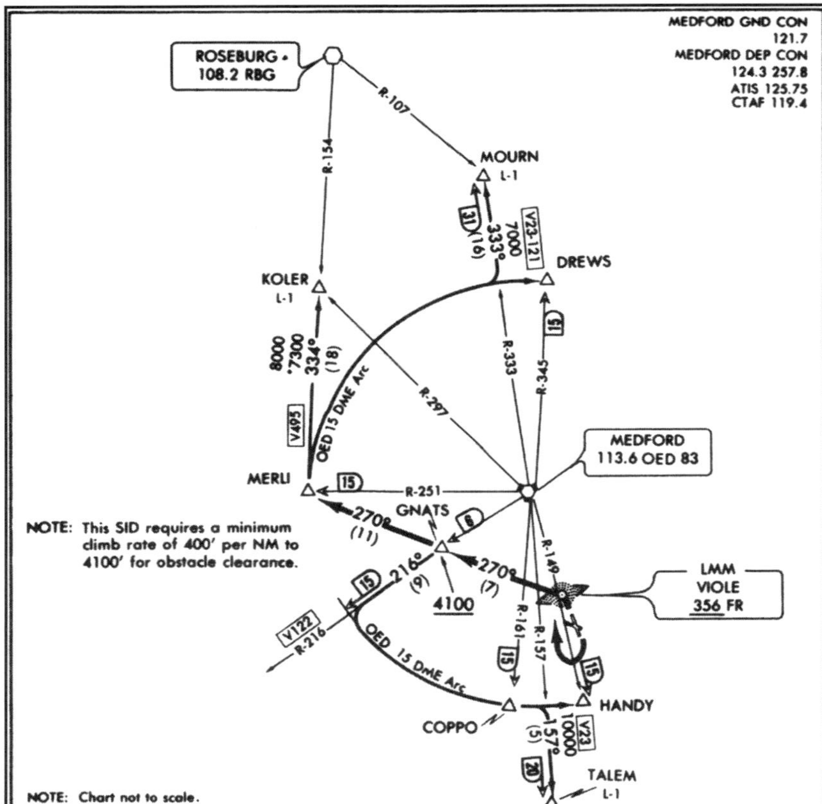

NOTE: This SID requires a minimum climb rate of 400′ per NM to 4100′ for obstacle clearance.

NOTE: Chart not to scale.

DEPARTURE ROUTE DESCRIPTION

Climb direct to the VIOLE ILS Middle Compass Locator (south take-off turn right), then climb on the 270° magnetic bearing from the LMM to GNATS INT, cross GNATS INT at or above 4100; thence via (transition) or (route).

COPPO TRANSITION (GNATS1.COPPO): Turn left via R-216 to 15 DME Fix, thence turn left via MEDFORD 15 DME Arc to COPPO DME Fix.

KOLER TRANSITION (GNATS1.KOLER): Continue via 270° magnetic bearing from the LMM to MERLI INT, turn right via ROSEBURG R-154 to KOLER INT.

(Continued on next page)

ELEV 1331

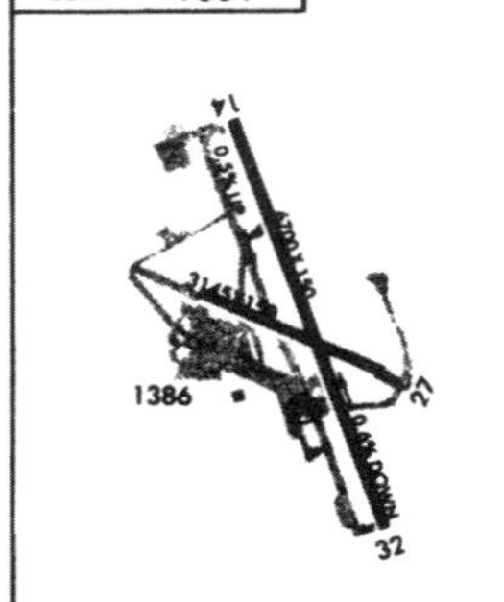

GNATS ONE DEPARTURE (GNATS1.GNATS) MEDFORD, OREGON MEDFORD-JACKSON CO

GNATS ONE DEPARTURE (GNATS1.GNATS)

MEDFORD-JACKSON CO
MEDFORD, OREGON

DEPARTURE ROUTE DESCRIPTION (Continued)

MOURN TRANSITION (GNATS1.MOURN): Continue via 270° magnetic bearing from the LMM to MERLI INT, turn right via MEDFORD 15 DME ARC to intercept V23-121 to MOURN INT.

DREWS TRANSITION (GNATS1.DREWS): Continue via 270° magnetic bearing from the LMM to MERLI INT, turn right via MEDFORD 15 DME ARC to DREWS INT.

TALEM TRANSITION (GNATS1.TALEM): Turn left via MEDFORD R-216 to 15 DME Fix thence turn left via MEDFORD 15 DME ARC to Intercept V23 to TALEM INT.

HANDY TRANSITION (GNATS1.HANDY): Turn left via MEDFORD R-216 to 15 DME Fix, thence turn left via MEDFORD 15 DME Arc to HANDY DME Fix.

GNATS ONE DEPARTURE (GNATS1.GNATS) MEDFORD, OREGON MEDFORD-JACKSON CO

§ **MEDFORD-JACKSON CO** (MFR) 3 N GMT–8(–7DT) 42°22′21″N 122°52′17″W **KLAMATH FALLS**
1331 B S4 **FUEL** 80, 100, 100LL, JET A1 + OX 1, 3 CFR Index B **H-1A, L-1A**
RWY 14-32: H6700X150 (ASPH-PFC) S-200, D-200, DT-400 HIRL .5% up S **IAP**
RWY 14: MALSR. Trees. **RWY 32:** REIL. VASI(V4L)—GA 3.0°TCH 49′. Road.
RWY 09-27: H3145X150 (ASPH) S-50, D-70, DT-108 MIRL
RWY 27: Road.
AIRPORT REMARKS: Attended continuously. CLOSED to unscheduled Part 121 air carriers operation, without prior approval, call 503-776-7222. Night refueling delay sunset-1500Z‡, ctc TOWER. Rwy 09-27 clsd to acft over 12,500 lbs GWT. Rwy 09/27 CLOSED when tower clsd. Rwy lgts 14/32 operate med ints when tower closed. ACTIVATE MALSR 14—119.4. Flocks of large waterfowl in vicinity Nov-May
COMMUNICATIONS: CTAF 119.4 **ATIS** 125.75 **UNICOM** 122.95
NORTH BEND FSS (OTH) LC 773-3256. NOTAM FILE MFR.
RCO 122.65 122.1R 113.6T (NORTH BEND FSS)
APP CON 124.3 (1400-0800Z‡) **DEP CON** 124.3 (1400-0800Z‡)
SEATTLE CENTER APP/DEP CON 125.3 (0800-1400Z‡)
TOWER 119.4 (1400-0800Z‡) **GND CON** 121.7
VFR ADVSY SVC ctc TOWER
RADIO AIDS TO NAVIGATION: NOTAM FILE OTH. VHF/DF ctc Medford TOWER
(H) ABVORTAC 113.6 OED Chan 83 42°28′47″N 122°54′43″W 146°6.1 NM to fld. 2080/19E
VORTAC unusable:
160°-165° beyond 35 NM below 8900′ 280°-345° beyond 30 NM below 6500′
198°-205° beyond 35 NM below 8500′ 345°-360° beyond 35 NM below 6800′
250°-280° beyond 25 NM below 6100′
PUMIE NDB (LOM) 373 MF 42°27′04″N 122°54′44″W 140°4.5 NM to fld. NOTAM FILE MFR
LOM unusable 150°-165° and 260°-265° beyond 5 miles.
VIOLE NDB (LMM) 356 FR 42°23′22″N 122°52′47″W 140°0.5 NM to fld. NOTAM FILE MFR
LMM unusable 305°-335° beyond 10 NM. all altitudes
ILS/DME 110.3 I-MFR Chan 40 Rwy 14 LOM PUMIE NDB. LMM VIOLE NDB. ILS unmonitored when tower closed. Localizer unusable inside threshold.

Volume 4 – Questions 8.3 & 8.4

122

(PILOT NAV) (STAKK2.STAKK) 91038
STAKK TWO DEPARTURE SL-192 (FAA)

HELENA REGIONAL
HELENA, MONTANA

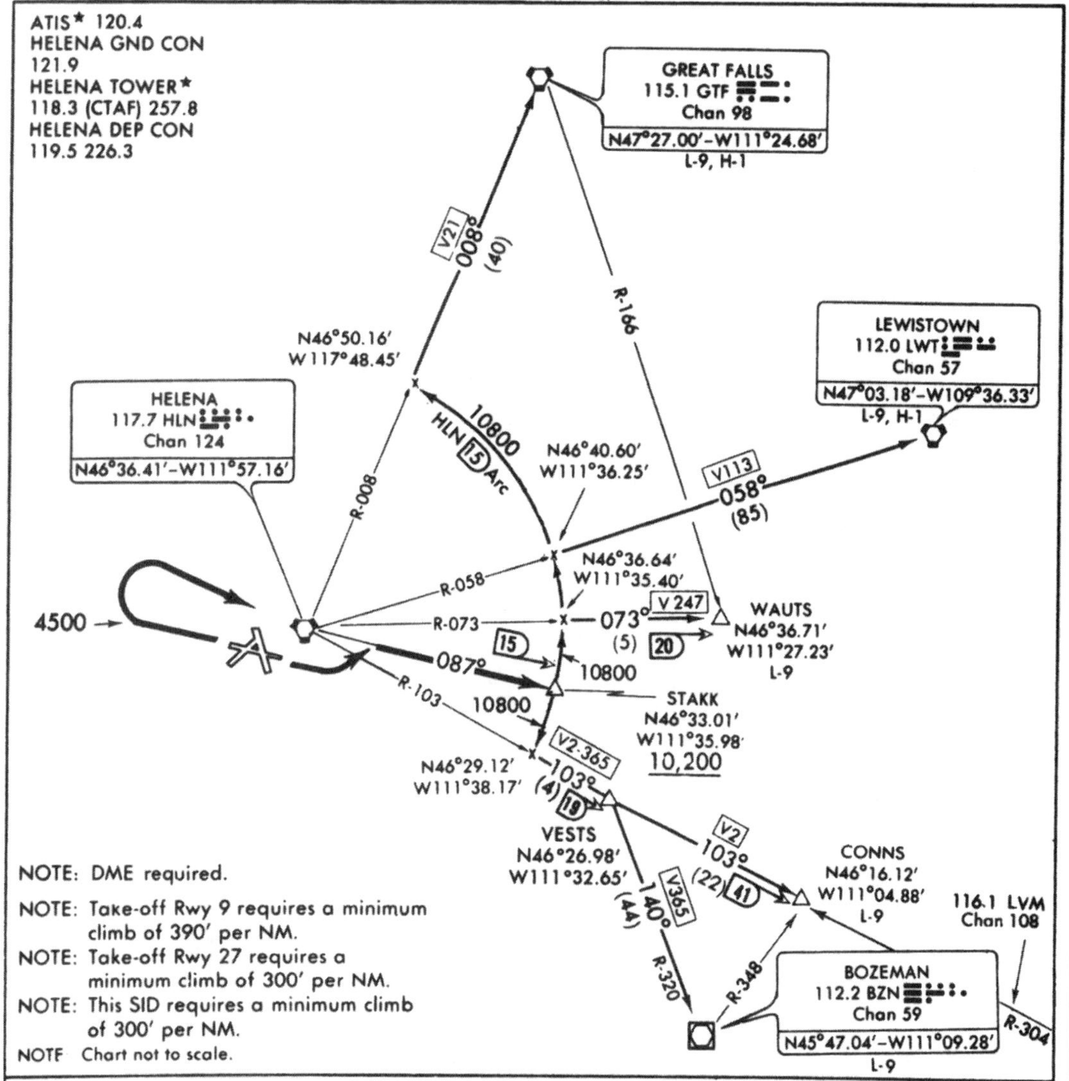

DEPARTURE ROUTE DESCRIPTION

TAKE-OFF RUNWAY 9: Climbing left turn to intercept HLN R-087 to cross STAKK INT at or above 10,200'. Thence via transition.

TAKE-OFF RUNWAY 27: Climb on runway heading to 4500' then climbing right turn direct HLN VORTAC. Continue climb on HLN R-087 to cross STAKK INT at or above 10,200'. Thence via transition.

BOZEMAN TRANSITION (STAKK2.BZN): From over STAKK INT via 15 DME Arc and V365 to BZN VOR/DME.

CONNS TRANSITION (STAKK2.CONNS): From over STAKK INT via 15 DME Arc and V2 to CONNS INT.

GREAT FALLS TRANSITION (STAKK2.GTF): From over STAKK INT via 15 DME Arc and V21 to GTF VORTAC.

LEWISTOWN TRANSITION (STAKK2.LWT): From over STAKK INT via 15 DME Arc and V113 to LWT VORTAC.

WAUTS TRANSITION (STAKK2.WAUTS): From over STAKK INT via 15 DME Arc and V247 to WAUTS INT.

STAKK TWO DEPARTURE
(PILOT NAV) (STAKK2.STAKK)

HELENA, MONTANA
HELENA REGIONAL

Volume 4 – Question 9.1

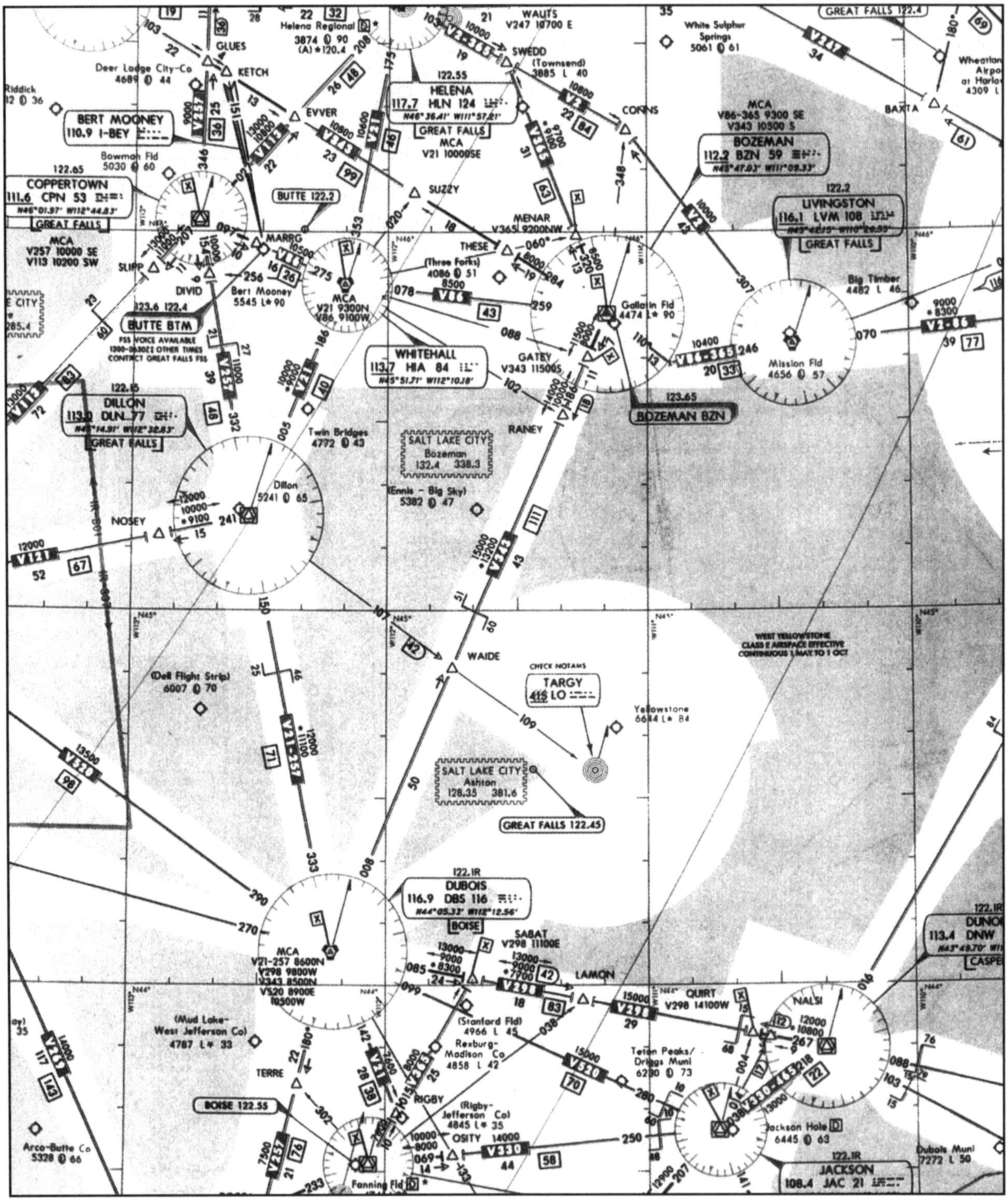

Volume 4 – Question 9.3

Amdt 29 91206

ILS RWY 6

AL-460 (FAA)

WINDSOR LOCKS/BRADLEY INTL (BDL)
WINDSOR LOCKS, CONNECTICUT

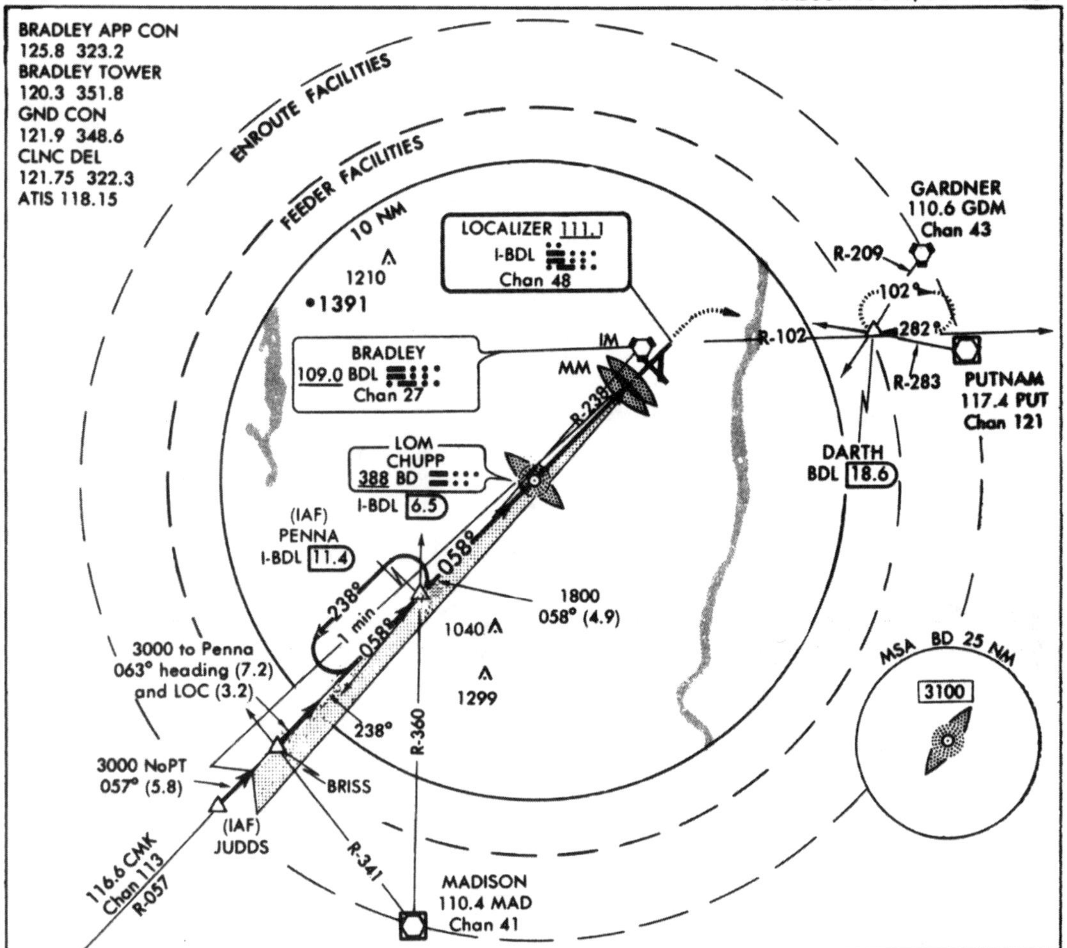

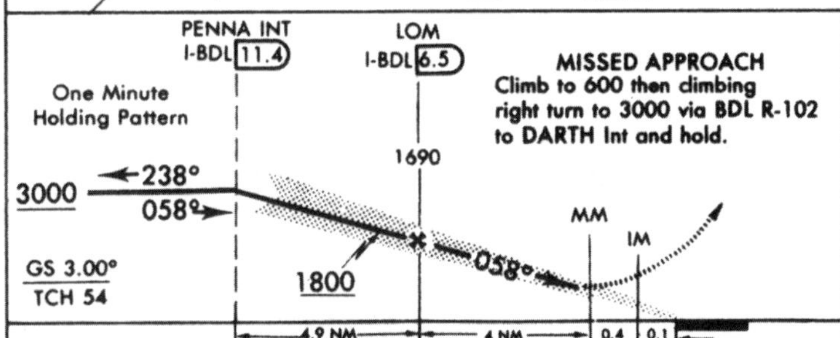

MISSED APPROACH
Climb to 600 then climbing right turn to 3000 via BDL R-102 to DARTH Int and hold.

ELEV 174

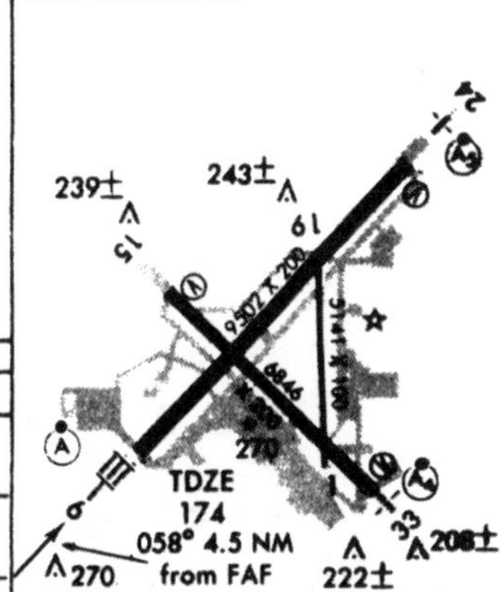

TDZ/CL Rwy 6
HIRL Rwys 6-24 and 15-33
MIRL Rwy 1-19
REIL Rwy 15

CATEGORY	A	B	C	D
S-ILS 6	374/18 200 (200-½)			374/20 200 (200-½)
S-LOC 6	760/24 586 (600-½)		760/50 586 (600-1)	760/60 586 (600-1¼)
CIRCLING	760-1 586 (600-1)		760-1½ 586 (600-1½)	1060-3 886 (900-3)

FAF to MAP 4.5 NM					
Knots	60	90	120	150	180
Min:Sec	4:30	3:00	2:15	1:48	1:30

ILS RWY 6

41°56'N-72°41'W

WINDSOR LOCKS, CONNECTICUT
WINDSOR LOCKS/BRADLEY INTL (BDL)

261

Volume 4 – Question 10.1

Orig 94230 AL-768 (FAA)

VOR/DME RNAV or GPS RWY 33

DALLAS/ADDISON (ADS)
DALLAS, TEXAS

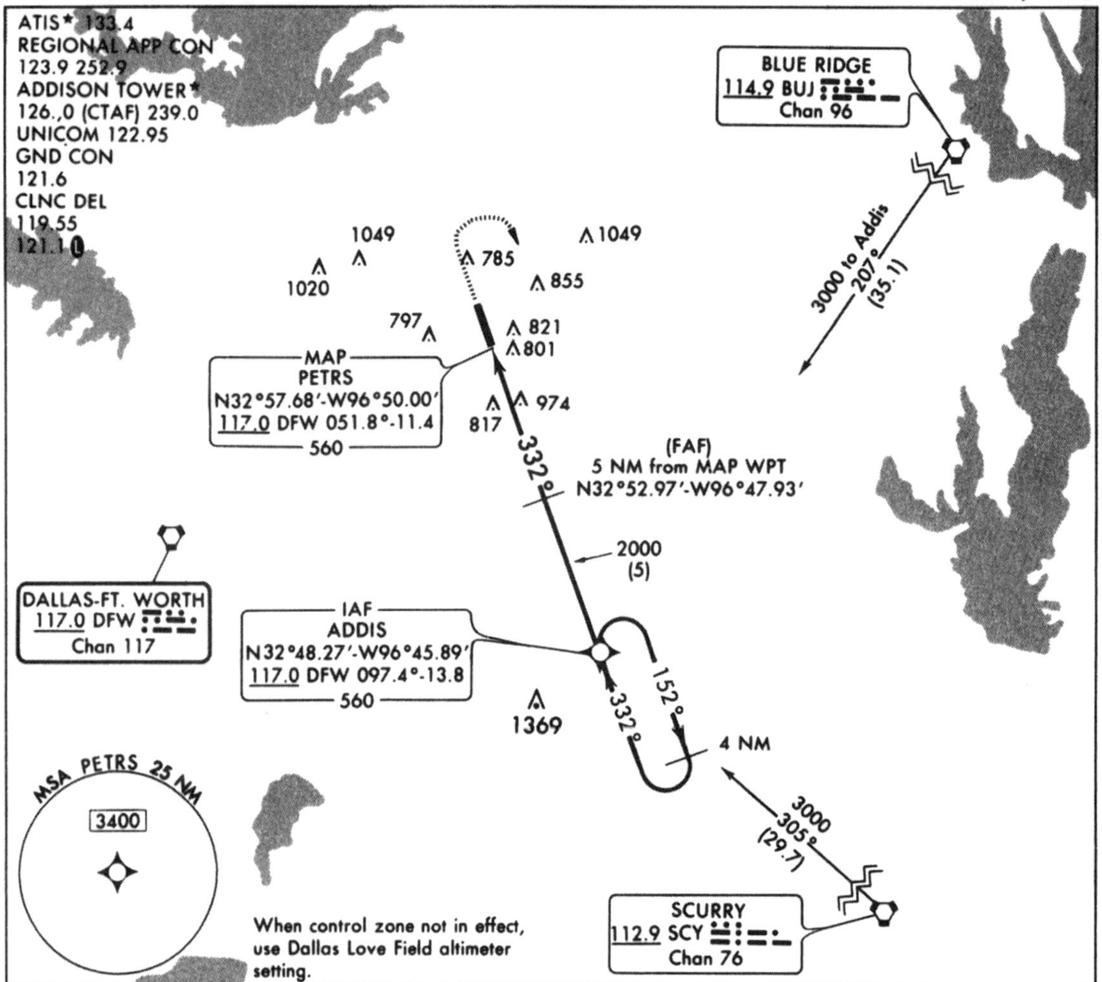

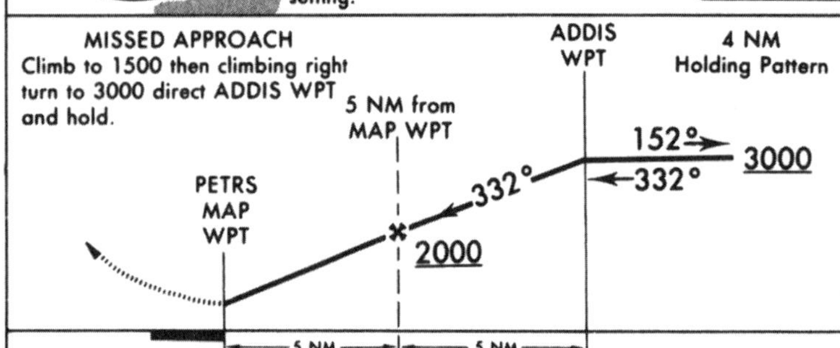

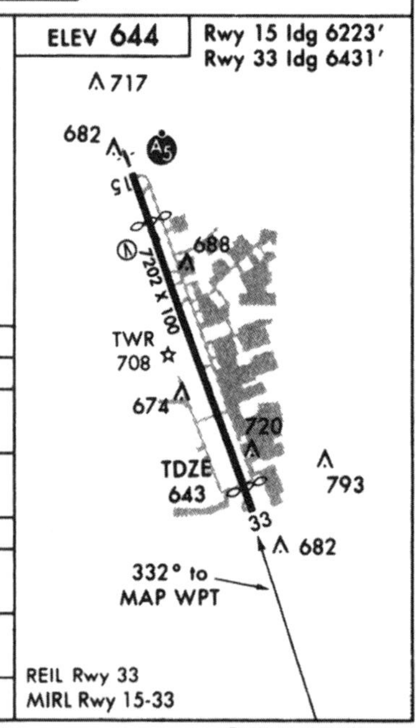

CATEGORY	A	B	C	D
S-33	1240-1 597 (600-1)		1240-1½ 597 (600-1½)	1240-1¾ 597 (600-1¾)
CIRCLING	1240-1 597 (600-1)		1240-1½ 597 (600-1½)	1280-2 637 (700-2)
DALLAS LOVE FIELD ALTIMETER SETTING MINIMUMS				
S-33	1300-1 657 (700-1)		1300-1¾ 657 (700-1¾)	1300-2 657 (700-2)
CIRCLING	1300-1 657 (700-1)		1300-1¾ 657 (700-1¾)	1360-2¼ 717 (800-2¼)

VOR/DME RNAV or GPS RWY 33

32°58'N-96°50'W

DALLAS, TEXAS
DALLAS/ADDISON (ADS)

Volume 4 – Question 10.2

91094 SL-6039 (FAA)

BLUE RIDGE THREE ARRIVAL (BUJ.BUJ3) DALLAS-FT. WORTH, TEXAS

ARRIVAL DESCRIPTION

FORT SMITH TRANSITION (FSM.BUJ3): From over FSM VORTAC via FSM R-213 and BUJ R-031 to BUJ VORTAC. Thence . . . .
LITTLE ROCK TRANSITION (LIT.BUJ3): From over LIT VORTAC via LIT R-244 and BUJ R-059 to BUJ VORTAC. Thence . . . .
TEXARKANA TRANSITION (TXK.BUJ3): From over TXK VORTAC via TXK R-272 and BUJ R-059 to BUJ VORTAC. Thence . . . .
TULSA TRANSITION (TUL.BUJ3): From over TUL VORTAC via TUL R-158 and BUJ R-031 to BUJ VORTAC. Thence . . . .
TURBOJETS LANDING DALLAS-FT WORTH INTL: (Landing South): From over BUJ VORTAC via BUJ R-230 to HAMAK INT. Expect vectors at BATON INT. (Landing North): From over BUJ VORTAC via BUJ R-230 to HAMAK INT, thence heading 170° for vector to final approach course.
NON-TURBOJETS LANDING DALLAS-FT WORTH INTL: (Landing South): From over BUJ VORTAC via BUJ R-230 to HAMAK INT. Expect vectors at BATON INT. (Landing North): From over BUJ VORTAC via BUJ R-215 to WEDER INT. Expect vectors to final approach course.
ALL AIRCRAFT LANDING DALLAS-LOVE FIELD, ADDISON, REDBIRD, NAS DALLAS, and PHIL L. HUDSON: (Landing South/North): From over BUJ VORTAC via BUJ R-215 to WEDER INT. Expect vectors to final approach course.
ALL AIRCRAFT LANDING MEACHAM, CARSWELL AFB, ALLIANCE, ARLINGTON, DENTON and FT. WORTH SPINKS: (Landing South/North): From over BUJ VORTAC via BUJ R-260 to KORKS INT. Expect vectors to final approach course.

Volume 4 – Question 10.3

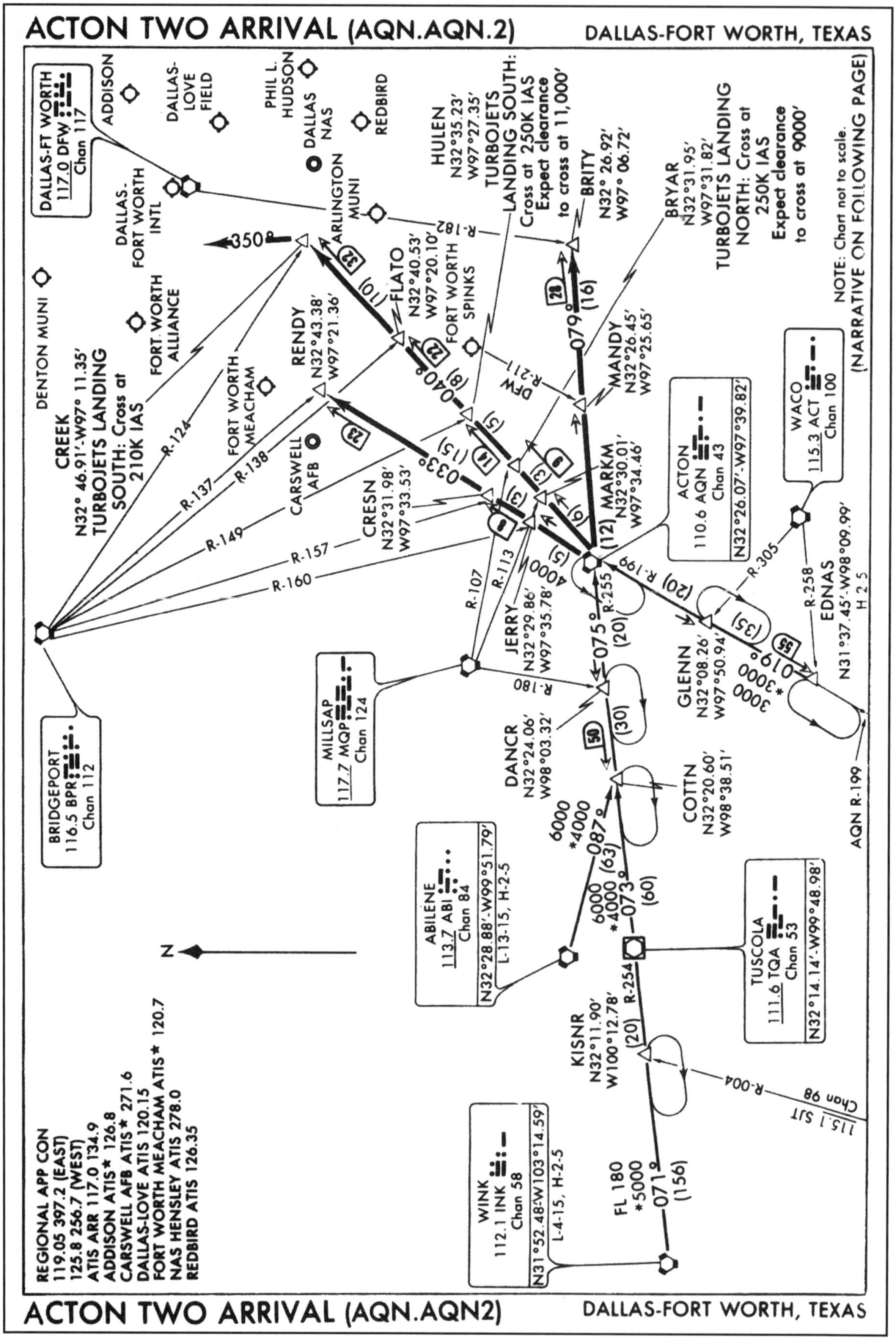

Volume 4 – Question 10.4

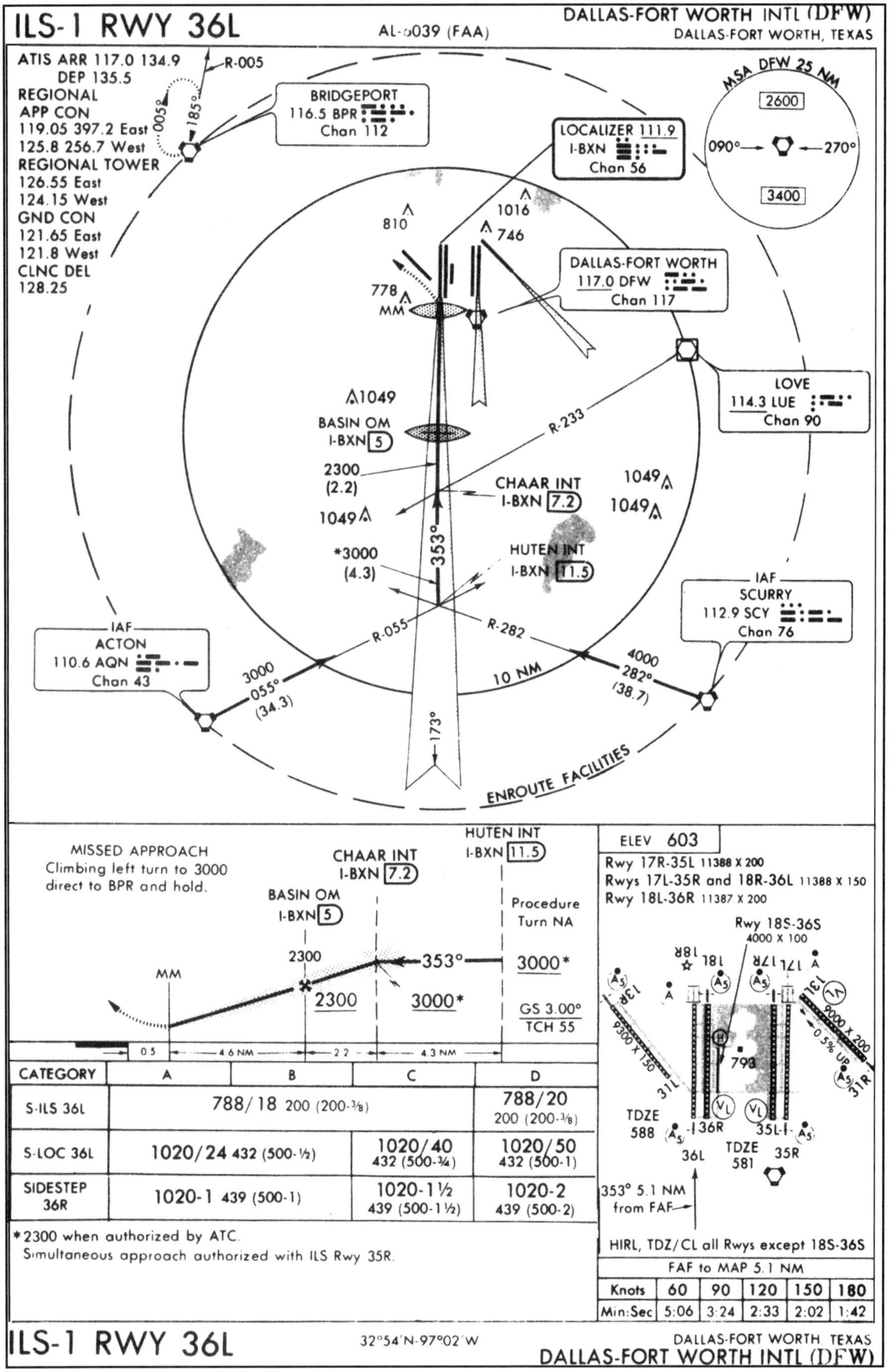

Volume 4 – Question 10.5

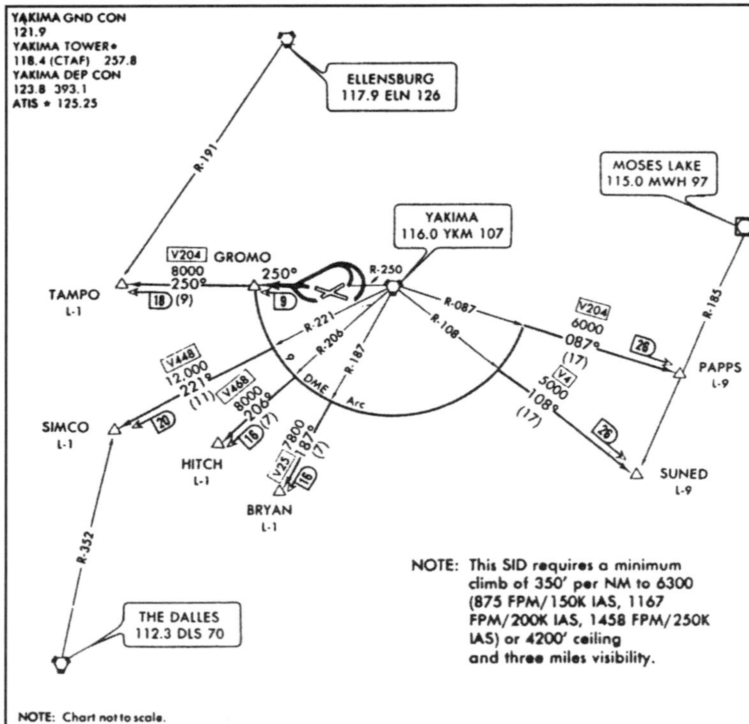

DEPARTURE ROUTE DESCRIPTION

TAKE-OFF RUNWAY 22: Turn right thence . . . .
TAKE-OFF RUNWAY 4, 9: Turn left thence . . . .
TAKE-OFF RUNWAY 27: Climb runway heading thence . . . .

. . . .intercept and proceed via YKM R-250 to the GROMO DME FIX (YKM R-250/9), thence via (assigned transition).
TAMPO TRANSITION (GROMO2.TAMPO): From GROMO DME Fix via V204 to TAMPO INT.
NOTE: Climb in holding pattern east of TAMPO INT, right turns, 250° inbound to 8000′ before proceeding on course.

(Continued on next page)

ELEV 1095

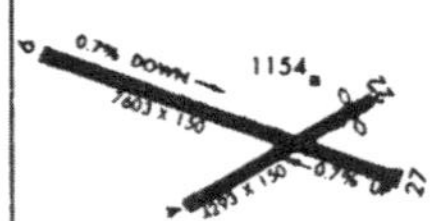

GROMO TWO DEPARTURE (PILOT NAV)
(GROMO2.GROMO)
YAKIMA, WASHINGTON
YAKIMA AIR TERMINAL

(GROMO2.GROMO)
GROMO TWO DEPARTURE (PILOT NAV)
YAKIMA AIR TERMINAL
YAKIMA, WASHINGTON

DEPARTURE ROUTE DESCRIPTION
(Continued)

SIMCO TRANSITION (GROMO2.SIMCO): From over GROMO DME Fix via 9 DME Arc and V448 to SIMCO INT. NOTE: Climb in holding pattern NE of SIMCO INT, right turns, 221° inbound to 12000 feet before proceeding on course.
HITCH TRANSITION (GROMO2.HITCH): From over GROMO DME Fix via 9 DME Arc and V468 to HITCH DME Fix. NOTE: Climb in holding pattern NE of HITCH DME Fix, right turns, 206° inbound to 8000 feet before proceeding on course.
BRYAN TRANSITION (GROMO2.BRYAN): From over GROMO DME Fix via 9 DME Arc and V25 to BRYAN DME Fix.
SUNED TRANSITION (GROMO2.SUNED): From over GROMO DME Fix via 9 DME Arc and V4 to SUNED INT.
PAPPS TRANSITION (GROMO2.PAPPS): From over GROMO DME Fix via 9 DME Arc and V204 to PAPPS INT.

GROMO TWO DEPARTURE (PILOT NAV)
(GROMO2.GROMO)
YAKIMA, WASHINGTON
YAKIMA AIR TERMINAL

§ **YAKIMA AIR TERMINAL** (YKM) 3 S GMT−8(−7DT) 46°34′06″N 120°32′33″W — SEATTLE H-1A, L-1C, 9A IAP
1095 B S4 FUEL 80, 100, 100LL, JET A OX 1, 3 CFR Index A
RWY 09-27: H7603X150 (ASPH-PFC) S-95, D-110, DT-175 HIRL .6% up NW
RWY 09: VASI(V4L)—GA 3.0° TCH 50′. Antenna. RWY 27: MALSR. Trees.
RWY 04-22: H4293X150 (ASPH) S-70, D-80, DT-120 MIRL
RWY 04: Trees. RWY 22: VASI(V2L)—GA 3.03° TCH 61′. Thld dsplcd 469′. Tree.
AIRPORT REMARKS: Attended continuously. CLOSED 0600-1400Z‡ to Part 121 non-scheduled operations and Part 121 scheduled charters operations without prior approval, call (509) 575-6150/6014. Bird hazard. ACTIVATE MALSR Rwy 27—118.4.
COMMUNICATIONS: CTAF 118.4 ATIS 125.25 (1400-0600Z‡) UNICOM 122.95
SEATTLE FSS (SEA) LC 965-1840 NOTAM FILE YKM
RCO 122.5 122.1R 116.0T (SEATTLE FSS)
APP/DEP CON 123.8 (1400-0600Z‡) SEATTLE CENTER APP/DEP CON 132.6 (0600-1400Z‡)
TOWER 118.4 (1400-0600Z‡) GND CON 121.9
VFR ADVSY SVC 123.8
RADIO AIDS TO NAVIGATION:
(H) VORTAC 116.0 YKM Chan 107 46°34′14″N 120°26′36″W 244° 3.6 NM to fld. 980/21E
DME portion unusable:

350°-065° beyond 20 NM below 9000′	025°-065° beyond 15 NM below 8000′
115°-135° beyond 25 NM below 7000′	135°-150° beyond 30 NM below 7500′
195°-230° beyond 22 NM below 8000′	300°-325° beyond 29 NM below 9000′
325°-335° beyond 20 NM below 9000′	335°-350° beyond 25 NM below 10,000′

VOR portion unusable:

350°-080° beyond 25 NM below 9000′	025°-035° beyond 5 NM below 10,000′
080°-105° beyond 35 NM below 6000′	105°-107° beyond 25 NM below 6000′
109°-135° beyond 25 NM below 6000′	135°-180° beyond 30 NM below 7500′
195°-225° beyond 30 NM below 8500′	305°-335° beyond 30 NM below 9000′

DONNY NDB (LOM) 371 YK 46°31′33″N 120°22′16″W 269° 7 NM to fld
ILS 110.1 I-YKM Rwy 27 LOM DONNY NDB. ILS unmonitored when tower closed. Localizer unusable inside threshold.

Volume 4 – Question 10.6

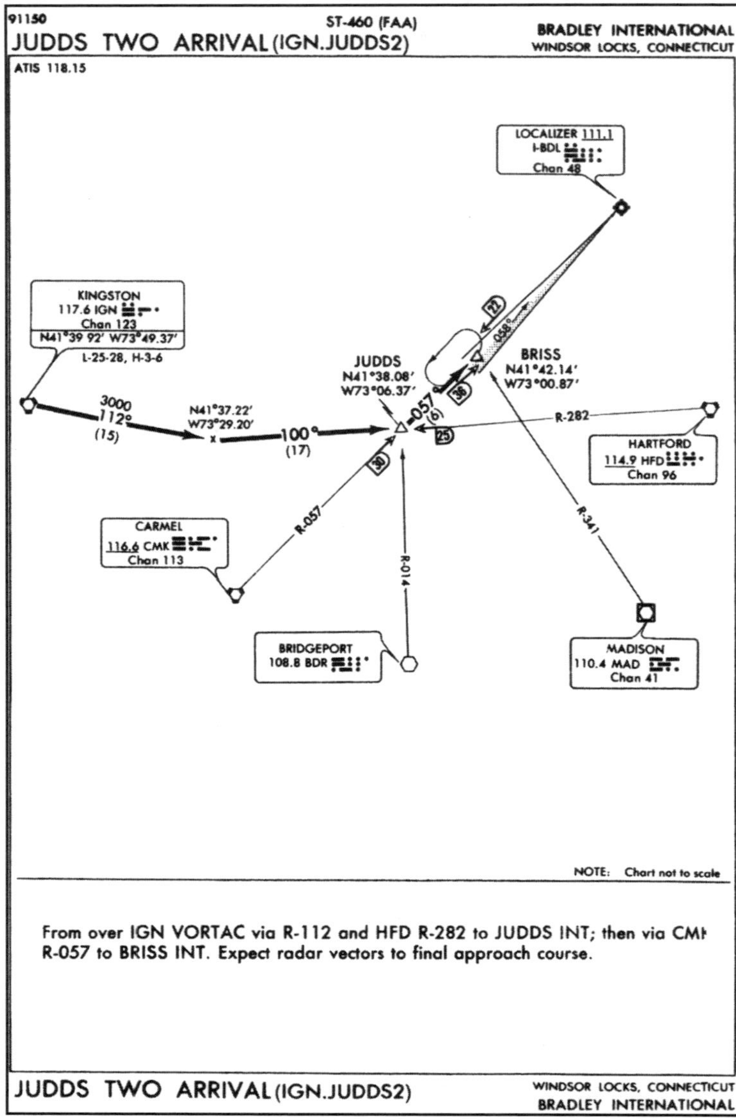

CONNECTICUT

WINDSOR LOCKS

BRADLEY INTL (BDL) 3 W UTC–5(–4DT) 41°56'20"N 72°41'01"W **NEW YORK**
174 B S4 **FUEL** 100LL, JET A OX 1, 2, 3, 4 TPA—See Remarks **H-3D, 6J, L-25C, 28 IAP**
LRA ARFF Index D

RWY 06-24: H9502X200 (ASPH-GRVD) S-200, D-200, DT-350,DDT-710 HIRL CL
RWY 06: ALSF2 TDZ. Trees. **RWY 24:** MALSR. VASI(V4L)—GA 3.0°TCH 56'.
RWY 15-33: H6846X200 (ASPH) S-200, D-200, DT-350 HIRL
RWY 15: REIL. VASI(V4L)—GA 3.5°TCH 59'. Trees. **RWY 33:** MALSF. VASI(V4R)—GA 3.0°TCH 59'. Trees
RWY 01-19: H5141X100 (ASPH) S-60, D-190, DT-328 MIRL
RWY 01: Building. **RWY 19:** Trees.

AIRPORT REMARKS: Attended continuously. Rwy 01–19 restricted to ldg and tkf with maximum tkf gross weight of 73,000 pounds. This restriction does not apply to acft emergency. Numerous birds frequently on or in vicinity or arpt. Portions of taxiway Alpha not visible from tower. TPA—1174(1000) light acft, 1874(1700) heavy acft. Landing fee for business, corporate and revenue producing aircraft. 24 hours ARFF level D svc avbl. 24 hours PPR for unscheduled air carrier ops with more than 30 passenger seats call arpt manager 203–627–3001/3008. This does not include delayed regularly schedule air carrier ops or diversions. Rwy 15 REIL out of svc indefinitely. Flight Notification Service (ADCUS) available. NOTE: See SPECIAL NOTICE—Simultaneous Operations on Intersecting Runways.

WEATHER DATA SOURCES: LLWAS.

COMMUNICATIONS: ATIS 118.15 **UNICOM** 122.95
BRIDGEPORT FSS (BDR) TF 1–800–WX–BRIEF. NOTAM FILE BDL.
WINDSOR LOCKS RCO 122.3 (BRIDGEPORT FSS)
Ⓡ **BRADLEY APP CON** 125.8 (within 20 miles)
Ⓡ **BRADLEY DEP CON** 121.05 (South) 125.35 (North and West) 123.95 (Northeast)
TOWER 120.3 **GND CON** 121.9 **CLNC DEL** 121.75
ARSA ctc **APP CON**

RADIO AIDS TO NAVIGATION: NOTAM FILE BDL.
(T) **VORTACW** 109.0 BDL Chan 27 41°56'27"N 72°41'21"W at fld. 165/14W.
VOR portion unusable 090°–103° beyond 24 NM below 5000' 104°–170° beyond 10 NM below 6000' 260°–290° beyond 15 NM below 6000'.
DME portion unusable:
040°–085° beyond 13 NM below 2000'
130°–150° beyond 10 NM below 3000'
170°–195° beyond 14 NM below 3000'
250°–290° beyond 18 NM below 6000'
CHUPP NDB (LOM) 388 BD 41°52'38"N 72°46'00"W 058° 5.2 NM to fld.
ILS/DME 111.1 I-BDL Chan 48 Rwy 06. LOM CHUPP NDB.
ILS/DME 108.55 I-IKX Chan 22Y Rwy 33
ILS/DME 111.1 I-MYQ Chan 48 Rwy 24

Volume 4 – Question 10.7

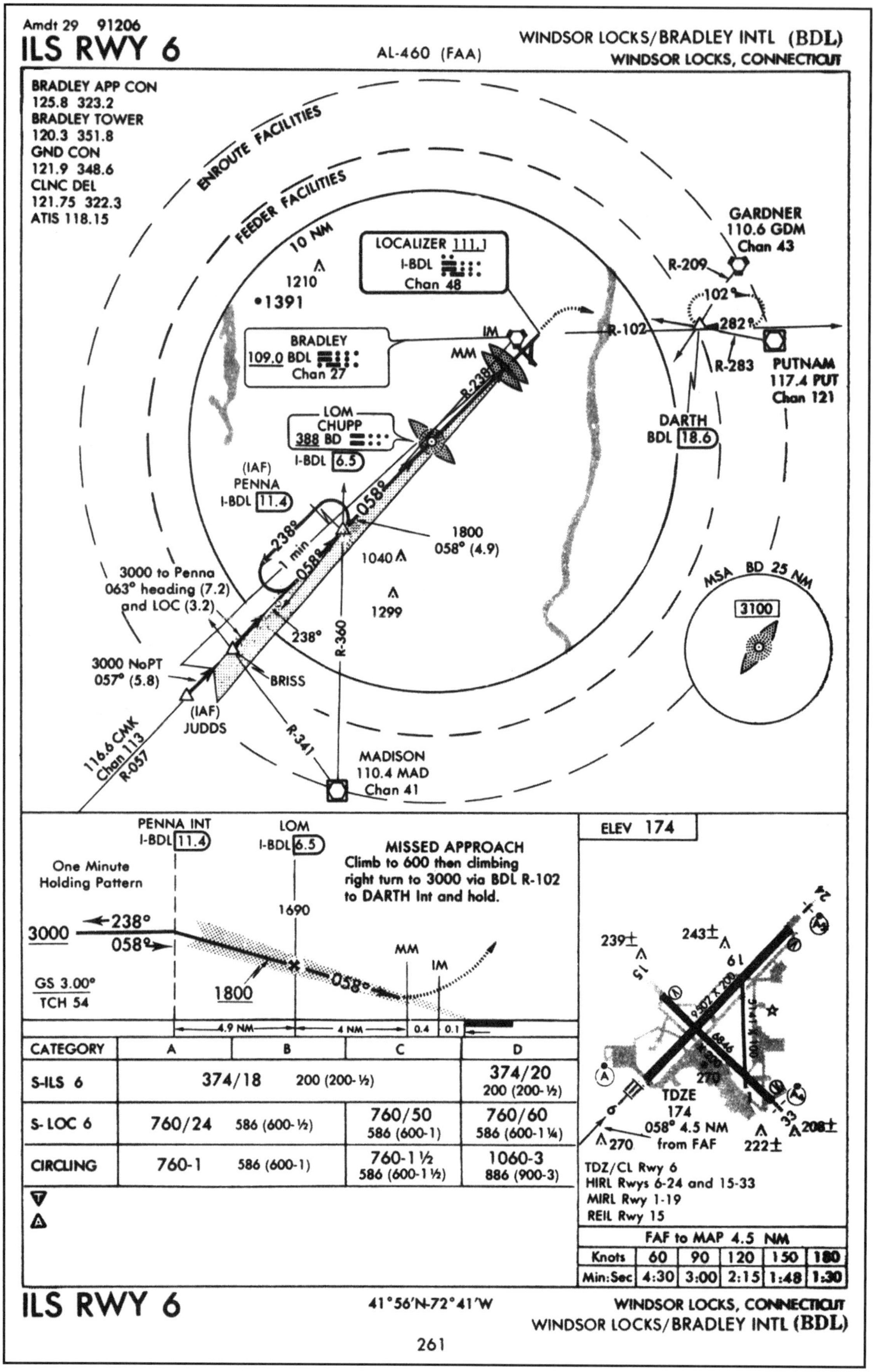

Volume 4 – Question 10.7

INOP COMPONENTS
00279

INOPERATIVE COMPONENTS OR VISUAL AIDS TABLE

Landing minimums published on instrument approach procedure charts are based upon full operation of all components and visual aids associated with the particular instrument approach chart being used. Higher minimums are required with inoperative components or visual aids as indicated below. If more than one component is inoperative, each minimum is raised to the highest minimum required by any single component that is inoperative. ILS glide slope inoperative minimums are published on the instrument approach charts as localizer minimums. This table may be amended by notes on the approach chart. Such notes apply only to the particular approach category(ies) as stated. See legend page for description of components indicated below.

(1) ILS, MLS, and PAR

Inoperative Component or Aid	Approach Category	Increase Visibility
ALSF 1 & 2, MALSR, & SSALR	ABCD	1/4 mile

(2) ILS with visibility minimum of 1,800 RVR

ALSF 1 & 2, MALSR, & SSALR	ABCD	To 4000 RVR
TDZL RCLS	ABCD	To 2400 RVR
RVR	ABCD	To 1/2 mile

(3) VOR,VOR/DME, VORTAC, VOR (TAC), VOR/DME (TAC), LOC, LOC/DME, LDA, LDA/DME, SDF, SDF/DME, GPS, RNAV,and ASR

Inoperative Visual Aid	Approach Category	Increase Visibility
ALSF 1 & 2, MALSR, & SSALR	ABCD	1/2 mile
SSALS,MALS, & ODALS	ABC	1/4 mile

(4) NDB

ALSF 1 & 2, MALSR, & SSALR	C	1/2 mile
	ABD	1/4 mile
MALS, SSALS, ODALS	ABC	1/4 mile

CORRECTIONS, COMMENTS AND/OR PROCUREMENT

FOR CHARTING ERRORS CONTACT:
National Aeronautical Charting Office, FAA
N/ACC1, SSMC-4, Sta. #2335
1305 East-West Highway
Silver Spring, MD 20910-3281
Telephone Toll-Free (800) 626-3677
Internet/E-Mail: Aerochart@NOAA.GOV

FOR CHANGES, ADDITIONS, OR RECOMMENDATIONS ON PROCEDURAL ASPECTS:
Contact Federal Aviation Administration, ATA 110
800 Independence Avenue, SW
Washington, DC 20591
Telephone Toll Free (800) 457-6656

TO PURCHASE CHARTS CONTACT:
National Aeronautical Charting Office
FAA, N/ACC3
Distribution Division
Riverdale, MD 20737
Telephone Toll Free (800) 638-8972

Requests for the creation or revisions to Airport Diagrams should be in accordance with FAA Order 7910.4B.

INOP COMPONENTS
00279

Volume 5 - Weather for IFR

Thunderstorms

1) High relative humidity, unstable air from the base to high levels, and a lifting force to set the cycle in motion are necessary for thunderstorm formation. The life cycle of a thunderstorm progresses through three stages, though it must be remembered that these stages apply only to cells. A storm cloud usually contains many such cells, all in different stages of development.
   a) The Cumulus Stage -- Each thunderstorm begins as a cumulus cloud, though few cumulus clouds develop into thunderstorms. This stage is characterized by a prevailing updraft that commonly reaches a speed of 3,000 feet per minute, and is strongest at higher altitudes during the later parts of the stage. Temperatures within the storm at this stage are higher than the surrounding air. Water droplets remain quite small during this updraft, but cause an icing hazard as temperatures decrease.
   b) The Mature Stage -- The transition to this stage is marked by surface precipitation. As the water droplets increase in size, they slow the updraft and cause a downdraft to occur. Updrafts remain, and may increase in speed to over 6,000 fpm, while the downdraft can exceed 2,500 fpm. These opposite drafts occur in very close proximity to each other, and cause very severe shear.
   c) The Dissipating Stage -- This stage arrives when the downdrafts have occupied all of the storm except the very top, where updrafts still persist. In this stage, the anvil top becomes apparent and precipitation ceases.
2) Types of thunderstorms:
   a) Air Mass Thunderstorms -- These thunderstorms form as a result of movement of different air masses, usually as a result of surface heating. They tend to be weaker in intensity and shorter in duration, but are still extremely dangerous to aircraft.
   b) Frontal and Prefrontal Thunderstorms -- These occur along and ahead of fronts, and may be embedded in other clouds. Therefore, these present a special risk for pilots, since visual circumnavigation is impossible. Frontal thunderstorms tend to be larger in size, intensity, and duration than air mass thunderstorms.
3) Flying in the vicinity of thunderstorms is extremely dangerous, since they contain many elements that are detrimental to airplanes. Turbulence occurs not only in the thunderstorms, but around them, also. The AIM states that turbulence can occur as far away as twenty miles from a severe thunderstorm.

Turbulence

1) Turbulence Reporting Criteria:
   a) **Light Turbulence** -- Causes slight, erratic changes in attitude and altitude. Occupants may feel a slight strain against seat belts, and unsecured objects may move slightly. A light, rhythmic bumpiness is called Light Chop.
   b) **Moderate Turbulence** -- Changes in attitude and altitude occur, but the airplane remains in positive control at all times. Occupants feel a definite strain on seat belts, and unsecured objects are dislodged. Turbulence similar to light chop that is more intense, but does not cause appreciable changes in altitude or attitude should be reported as Moderate Chop.
   c) **Severe Turbulence** -- Causes large, abrupt changes in attitude and altitude, as well as a greatly varying indicated airspeed. May cause temporary loss of control. Occupants are forced violently against seat belts, and unsecured objects are tossed about.
   d) **Extreme Turbulence** -- Aircraft is violently tossed about, and control is practically impossible. May cause structural damage to aircraft.
2) Since the intensity of the effect of turbulence relies on the size of the airplane encountering the turbulence, it is necessary to give the type of aircraft when reporting turbulence. A small airplane will report turbulence as stronger than a large airplane.
3) Types of Turbulence:
   a) **Convective turbulence** is nearly always present on warm days and is most active when winds are light. It is especially heavy over mountains and high deserts in the west, and is strongest in the mid-afternoon.
   b) **Wind shear** is a change in wind speed and/or wind direction in a short distance resulting in a tearing or shearing effect, and occurs in both horizontal and vertical directions. This is especially hazardous near the surface.

4) An airplane's **maneuvering speed** is a mathematically calculated speed at which the airplane will stall as the limit load factor is reached. It is directly proportional to the weight of the airplane, and therefore increases as the weight increases. Most handbooks give the maneuvering speed for maximum gross weight. Since the maneuvering speed is lower than the cruising speed, the airplane should be slowed down when entering stronger turbulence.

Structural Icing

1) The effects of structural icing can be dangerous. Icing will cause lift for a given angle of attack to decrease, thrust to decrease, and drag and weight to increase. Therefore, in conditions of icing, the stall speed for the airplane will rise sharply.

2) Structural icing occurs when flying through areas containing supercooled water droplets, or droplets below freezing temperature, but still in liquid form. Therefore, visible moisture is necessary for icing to occur. The rate at which ice accumulates depends upon the amount of liquid water, the size of the supercooled drops, airspeed, and size and shape of the airfoil.

3) The most severe icing will take place in areas where there is mechanical lifting. The severity of ice is directly related to the stability of the air. Therefore, icing encountered in turbulence will likely be very severe.

4) There are two main types of icing, which may appear together as Mixed Ice.
   a) **Clear ice** occurs when supercooled droplets freeze slowly after contacting the airplane's surface. Since more of the trapped air escapes from the droplet after contact, clear ice is more solid and smoother than rime ice. Although it does not spoil the airflow as quickly as rime ice, it is much more difficult to remove due to its solidity.
   b) **Rime ice** occurs when supercooled droplets freeze instantaneously upon contact with the airplane's surface. This causes air to become trapped in the ice, and makes the surface rough and milky in color. Rime ice is easy to remove, but will cause the airflow to lose efficiency fairly rapidly because of its roughness. Rime ice is usually found in stratus clouds in ambient temperatures at freezing or below. Rime ice is possible at temperatures below -10° C.

5) Icing is often found in dynamically changing weather conditions. Therefore, PIREPs concerning icing are useful only if they are fairly recent. Also, icing reports are more useful when reported by airplanes with airspeeds similar to yours.

6) Test data indicate that ice, snow, or frost formations having a thickness and surface roughness similar to medium or course sandpaper on the leading edge and upper surface of a wing can reduce wing lift by as much as 30% and increase drag by 40%.

7) The roughness of frost spoils the smooth flow of air over the airfoil causing a slowing of the air. Even a small amount of frost can cause early airflow separation resulting in a loss of lift.

Weather Minimums

1) NOTE: Weather minimums for takeoff do not apply to not-for-hire flight operations. Instrument approach minimums do apply to all aircraft and operations.

2) ILS Approach Minimums:
   a) 1/2-mile visibility is required for most approaches, although some approaches require only 1800 feet of runway visual range.
   b) Decision height is usually 200 feet above the runway touchdown zone.

3) Non-precision Approach Minimums:
   a) VOR and Localizer Approaches:
      i) 1/2 mile visibility
      ii) Minimum Descent Altitudes (MDAs) from 400 to 600 feet AGL. The lowest possible MDA for a VOR approach is 250 feet AGL.
   b) NDB Approaches:
      i) 1/2 mile visibility
      ii) 350 foot MDA

4) **RVR** -- The Runway Visual Range is measured from the base of the approach end of the runway, and is reported in hundreds of feet. It represents the horizontal distance a pilot will see down the runway from the approach end. It is a horizontal measurement, not a slant measurement.

5) More accidents occur during approaches than during any other time in IFR flights, and data from these accidents suggests that descent past the MDA or DA was intentional, and was therefore caused by pilots pushing the approach to catch sight of the runway, or by pilots who believed the runway was in sight.

6) Accidents also occur often when multiple approaches are made, where the pilot will try progressively harder to make a successful approach, even though there is no visibility at the MDA.

Fog

1) **Radiation fog** is caused by cooling of the land surface, which is then commuted to the surrounding air. Radiation fog usually occurs on clear nights when the temperature – dew point spread is small, and winds are very light. It also forms or thickens right after sunrise.

2) **Advection fog** is a coastal fog, it occurs when warm, moist air moves over a relatively cold surface in moderate wind conditions. Heavier winds will lift the fog, forming a stratus cloud. It is very difficult to forecast when advection fog will clear.

3) **Upslope fog** is produced by air cooling as it expands moving up a slope. It will also form in moderate winds, as heavy winds will produce stratus or stratocumulus clouds instead of fog.

4) Note: Fog (FG) is reported only when the visibility is less than 5/8 S.M., otherwise mist (BR) is reported. Total obscurations of vertical visibility are reported in the format "VVhhh" where VV denotes vertical visibility and "hhh" is the vertical visibility in hundreds of feet. There is no provision in the METAR code to report partial obscurations.
   a) Example: 1/8SM FG VV006
      i) Horizontal visibility one eighth of a mile in fog, vertical visibility 600 feet.

Volume 6 - Advanced IFR

Supplemental Oxygen

1) According to 14 CFR §91.211, at cabin pressure altitudes above 12,500 feet MSL up to and including 14,000 feet MSL, the flight crew must be provided with and use supplemental oxygen for any part of the flight at those altitudes that is more than 30 minutes in duration. At cabin pressure altitudes above 14,000 feet MSL, supplemental oxygen is required for the flight crew during the entire flight time at those altitudes. Above 15,000 feet MSL, each occupant must be provided with supplemental oxygen for the duration of the flight above that altitude.

Area Navigation

1) RNAV, or Area navigation, provides enhanced navigational capability to the pilot. RNAV equipment can compute the airplane position, actual track, and ground speed, and then provide meaningful information relative to a route of flight selected by the pilot. Typical equipment will provide the pilot with distance, time, bearing, and crosstrack error relative to the selected "TO" or "active" waypoint and the selected route. Several navigational systems with different navigational performance characteristics are capable of providing area navigational functions. Present day RNAV includes INS, LORAN, VOR/DME, and GPS.

LORAN & GPS

1) LORAN, or Long Range Navigation, is based on time differentials between synchronized pulse signals from different ground transmitters. GPS, or Global Positioning System, uses travel time of signals from different satellites. Both systems give the same result--the position of the airplane in terms of latitude and longitude. This information, along with information on the destination or fixes along a route, is used by the computer in the on-board receiver to compute bearing to the next point in the flight plan, or the destination, as well as the actual track, ground speed, estimated time en route, and the desired track based on the flight plan. LORAN units can be approved for IFR en route and terminal navigation. GPS units can also be approved for approach navigation using either stand-alone GPS approaches, or overlays of existing non-precision approaches. With either system, the installation must be specifically approved by the FAA for IFR use.

2) VOR/DME RNAV, LORAN and GPS systems all show deviation from course in linear terms, where VOR deviation is shown in angular terms. For example, a one dot cross track error on a VOR indicates that the airplane is two degrees off course. The same error would indicate one mile off course when using one of the other navigational systems in en route flying. When GPS is used for an approach, the sensitivity of the cross track error indication increases as the airplane closes on the destination.

Horizontal Situation Indicator (HSI)

1) The Horizontal Situation Indicator is a combination of two instruments, the vertical azimuth card heading indicator, and the VOR/ILS indicator. The course deviation bar operates with a VOR/LOC navigation receiver to indicate left or right deviations from the course selected, and moves left or right in the same manner that the angular movement of a conventional VOR/LOC needle indicates deviation from course. Desired course is selected by rotating the course-indicating arrow in relation to the heading azimuth card with the course set knob. The triangle shows if course is to or from the station.

2) The HSI gives you a pictorial presentation, as if you were above the aircraft looking down, and displays the aircraft, course selected, and relative navaid location.

Distance Measuring Equipment (DME)

1) DME is used in conjunction with VORs and, on rare occasions, NDBs. As the name implies, DME will tell you distance from the appropriate station. The mileage readout is in nautical miles and is the direct distance from the aircraft to the DME ground facility. This is commonly referred to as the slant-range distance. The difference between a measured distance on the surface and the DME slant-range distance is known as slant-range error, and is smallest at low altitude and long range. The error is greatest when the aircraft is directly over the ground facility. To consider readings accurate, you should be at least one nautical mile from the station for each 1,000 feet above the station.

2) According to 14 CFR §91.205, an aircraft must be equipped with approved DME when operating at or above FL 240 when using VORs for navigation. If the DME fails above this altitude, the pilot shall notify ATC immediately, and then may continue operations at and above FL 240 to the next airport of intended landing at which repairs or replacement of the equipment can be made.
    a) NOTE: IFR approved GPS may be used as a substitute for DME, and satisfies the DME requirement of 14 CFR §91.205.

Volume 6 – Lebanon-Warren County RNAV Runway 18

LEBANON, OHIO

AL-6485 (FAA)

APP CRS	Rwy ldg	TDZE	Apt Elev
189°	4092	897	898

RNAV RWY 18
LEBANON-WARREN COUNTY (I68)

NA Obtain local altimeter on CTAF; when not received use Dayton-Wright Brothers altimeter setting.
VDP NA with Dayton-Wright Brothers altimeter setting.

MISSED APPROACH: Climb to 2000 then climbing right turn to 3000 direct MIZZA WP and hold.

DAYTON APP CON	CLNC DEL	UNICOM
118.85 327.1	119.4	123.05 (CTAF) 123.5

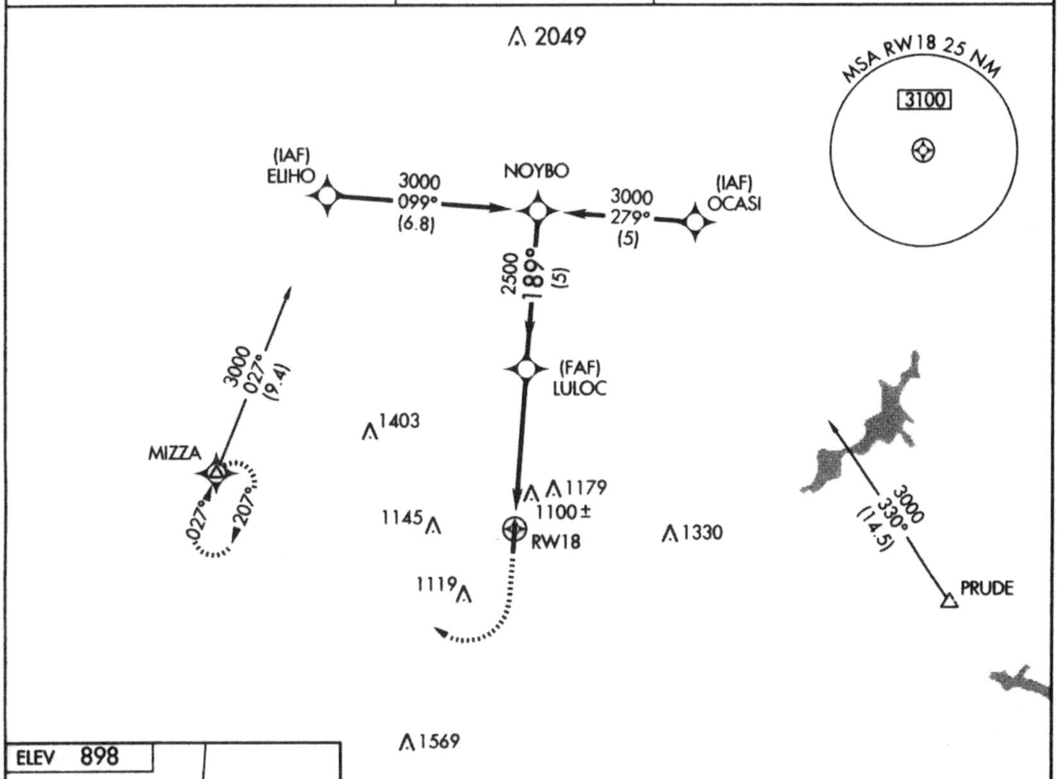

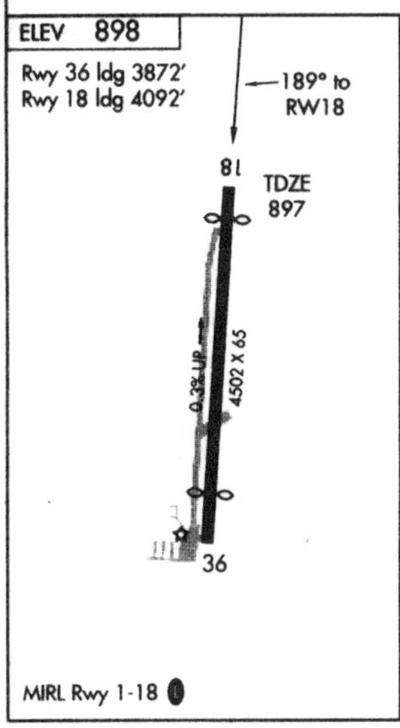

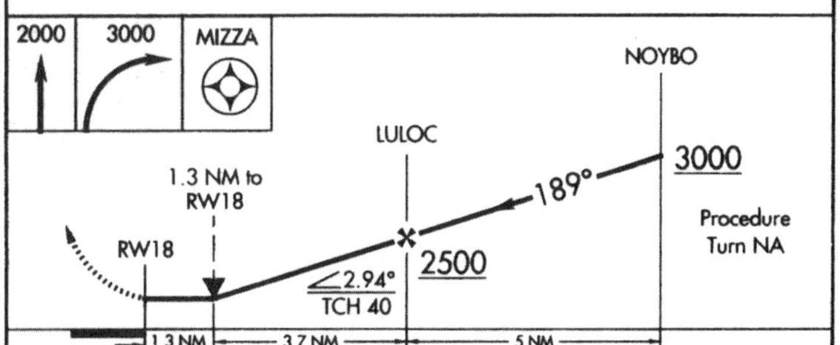

CATEGORY	A	B	C	D
LNAV MDA	1360-1 463 (500-1)		1360-1¼ 463 (500-1¼)	1360-1½ 463 (500-1½)
CIRCLING	1380-1 482 (500-1)	1400-1 502 (600-1)	1480-1½ 582 (600-1½)	1480-2 582 (600-2)
DAYTON-WRIGHT BROTHERS ALTIMETER SETTING MINIMUMS				
LNAV MDA	1380-1 483 (500-1)		1380-1¼ 483 (500-1¼)	1380-1½ 483 (500-1½)
CIRCLING	1420-1 522 (600-1)		1520-1¾ 622 (700-1¾)	1520-2 622 (700-2)

LEBANON, OHIO
Orig-A 00167

39°28'N - 84°15'W
309

LEBANON-WARREN COUNTY (I68)
RNAV RWY 18

Volume 6 – Batavia/Clermont County NDB or GPS Runway 22

Orig-B 00223

NDB or GPS RWY 22 AL-5955 (FAA) BATAVIA/ CLERMONT COUNTY (I69) BATAVIA, OHIO

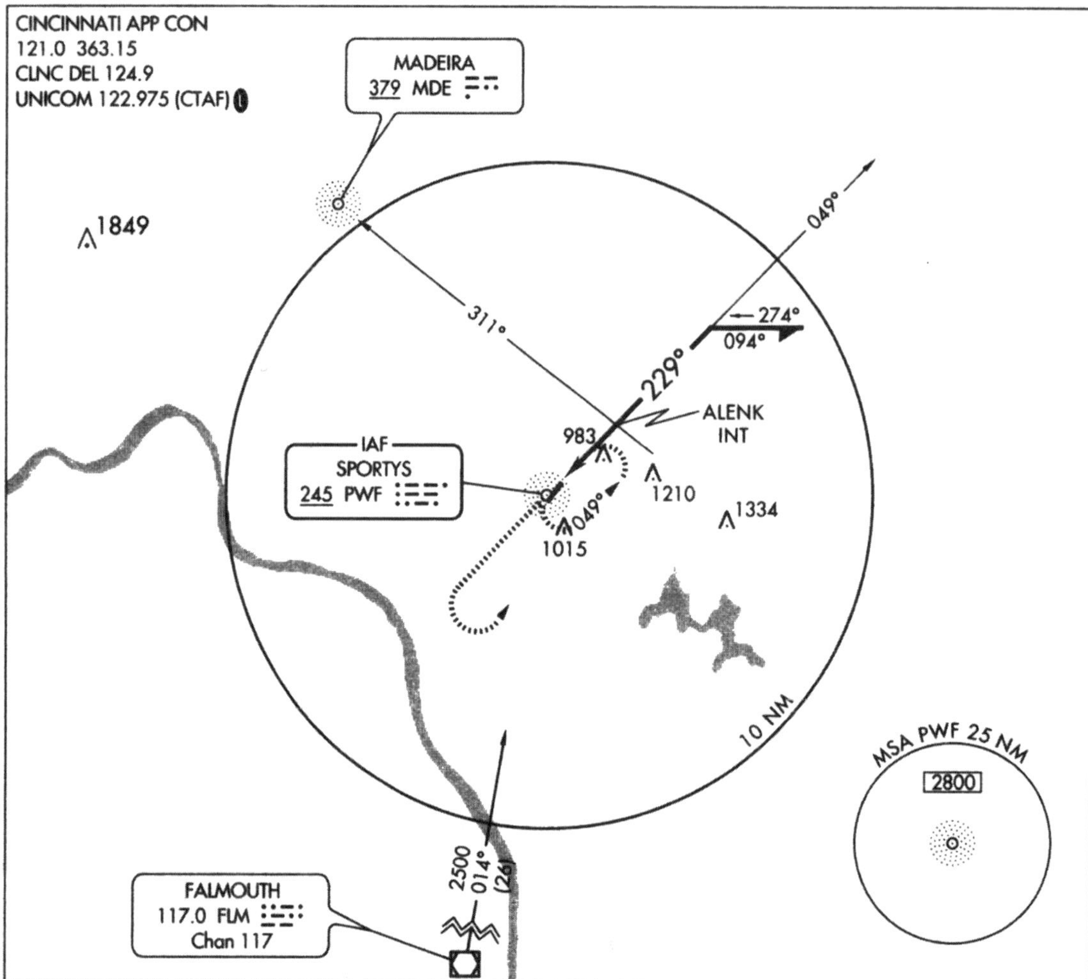

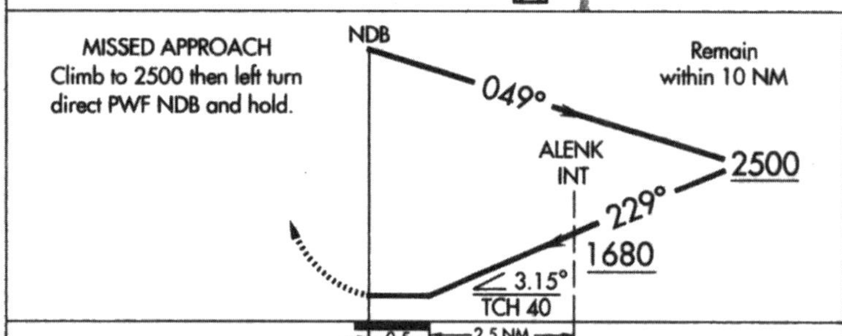

CATEGORY	A	B	C	D
S-22	1680-1 836 (900-1)	1680-1¼ 836 (900-1¼)	1680-2½ 836 (900-2½)	NA
CIRCLING	1680-1 832 (900-1)	1680-1¼ 832 (900-1¼)	1680-2½ 832 (900-2½)	NA
DUAL ADF MINIMUMS				
S-22	1380-1 536 (600-1)		1380-1½ 536 (600-1½)	NA
CIRCLING	1400-1 552 (600-1)		1400-1½ 552 (600-1½)	NA

Use Cincinnati Lunken Field altimeter setting. ▼ A NA

ELEV 848 Rwy 4 ldg 3174'

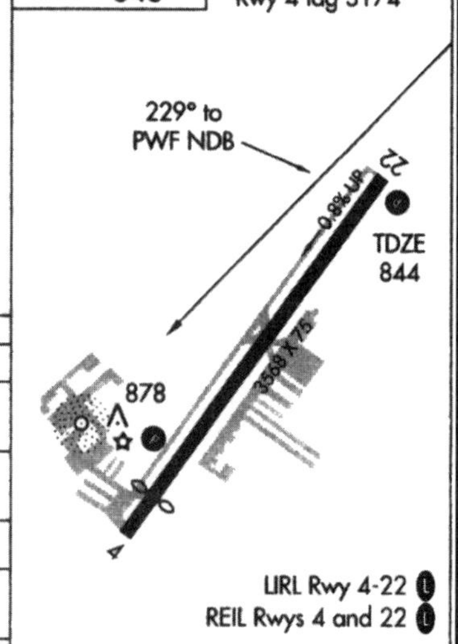

Knots	60	90	120	150	180
Min:Sec					

NDB or GPS RWY 22 39° 05'N - 84° 13'W BATAVIA, OHIO BATAVIA/ CLERMONT COUNTY (I69)

Orig-B 00223

Volume 6 – Question 10.5

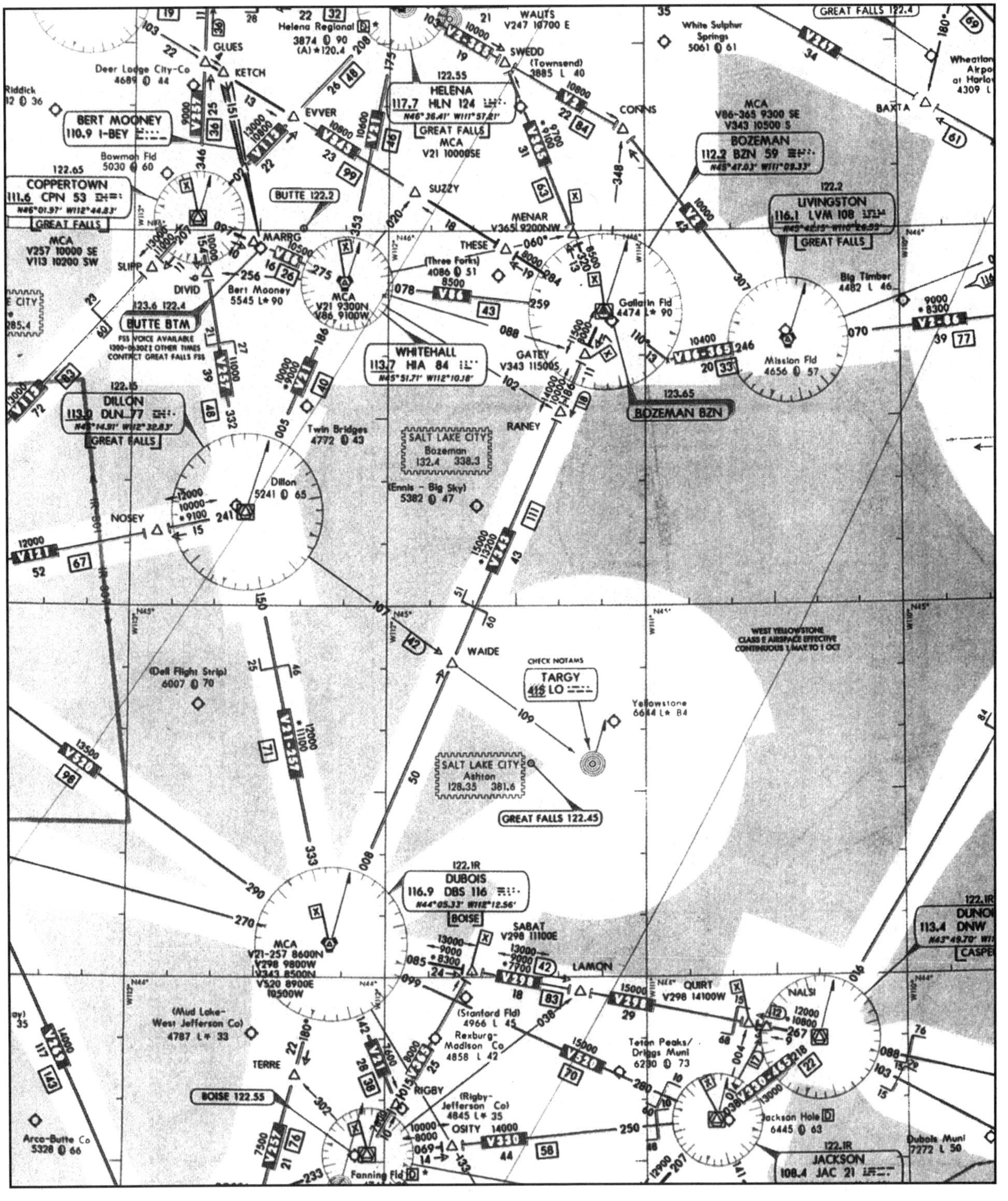

Volume 7 - FARs and Your Instrument Test

Special Clearances

1) **"VFR-On-Top" Clearance** -- an ATC authorization for an IFR aircraft to operate in VFR conditions at any appropriate VFR altitude above the MEA for the route. The "VFR-on-top" clearance must be requested by the pilot, and permits the pilot to select an altitude of his choice in lieu of an assigned altitude. However, the pilot still must report changes in altitude to ATC. On this clearance, ATC may or may not offer traffic advisories, and it is the sole responsibility of the pilot to see and avoid other aircraft.

2) **Cruise Clearance** -- Used in place of "maintain", "cruise" assigns a block of airspace from the minimum IFR altitude up to and including the altitude specified in the cruise clearance. The pilot may level off at any intermediate altitude within this block of airspace. Climb/descent within the block is to be made at the discretion of the pilot. However, once the pilot starts descent and verbally reports leaving an altitude in the block, he may not return to that altitude without additional ATC clearance.

3) **Abbreviated Clearance** -- given by ATC when an IFR flight plan can be approved with little or no revision. A controller will issue an abbreviated clearance by stating "cleared to (destination) as filed"; however, this clearance does not include approval of the requested altitude, only course. Altitude will always be given in the clearance. If a DP is assigned, the name of the DP and transition will be given.

4) **Clearances from Non-Towered Airports** -- If surface radio communication with ATC is possible, the clearance would be given over the radio. If no radio communication from ATC is possible, the clearance would be received over the telephone. Clearances given to airplanes departing from non-towered airports include time parameters for the takeoff; the earliest the airplane may take off is the release time, and it must have departed before the void time. If departure has not taken place before the void time, ATC must be notified as soon as possible, but no later than 30 minutes after the void time, when the search and rescue process will begin.

Composite Flight Plans

1) A Composite flight plan specifies VFR operation for one portion of flight and IFR for another portion, either of which may be first. The pilot must request the closing of the VFR portion of the flight plan regardless of the type of facility being communicated with.

VOR Equipment Checks

1) In order to conduct IFR flights, VOR receivers must have been operationally checked and found within the permissible indicated bearing error limits within the past 30 days. Each person making a VOR operational check must record the date, place, bearing error and then sign the entry. The entry should be made in the aircraft log or other record.

2) Use of an FAA-operated or approved test signal (VOT), or a specific point on the airport surface designated as a VOR system checkpoint. The maximum permissible bearing error for these methods is ±4°. With the receiver tuned to a VOT and the CDI centered, the selector should read either "360° from" or "180° to." If authorized in the A/FD, a VOT may be used while airborne.

3) Use of a specifically designated airborne checkpoint. The maximum permissible bearing error for this method is ±6°, and should be used only when a VOT or surface checkpoint or signal is unavailable.

4) For airplanes with dual system VORs installed, the test may be conducted by checking one system against the other. Both systems should be tuned to the same VOR ground facility and the indicated bearings to the station noted. The maximum permissible difference between the two indicated bearings is 4°. A dual VOR check may be performed either on the ground or while airborne.

5) NOTE: Each dot on a VOR receiver represents a 2° deviation. Also, when tuning to a VOR, if an identifier is absent, the station is undergoing routine maintenance, and while there may be a signal, it could give erroneous indications.

Federal Aviation Regulations

1) According to 14 CFR §61.51(g), instrument flight time may only be logged when the pilot operates the aircraft solely by reference to instruments, under actual or simulated instrument flight conditions. An authorized instructor may log instrument time when conducting instrument flight instruction in actual instrument flight conditions. Logbook entries must include location and type of each instrument approach accomplished, and the name of the safety pilot if required.

2) According to 14 CFR §61.57(c), no pilot may act as pilot in command under IFR conditions or conditions less than VFR unless the pilot has completed the following in the previous six calendar months:
   a) At least six instrument approaches, holding procedures, intercepting and tracking courses through the use of navigation systems; or
   b) Passed an instrument proficiency check as prescribed in 14 CFR §61.57(d).

3) Any pilot who does not meet these recent experience requirements cannot serve as pilot in command under IFR, nor in weather less than prescribed VFR minimums. Any pilot who has not met these requirements for a period of less than six months may regain currency by completing the items listed in 2)a) above or by completing an instrument proficiency check. Any pilot who has not met the recent experience requirements for a period of six months or more may only regain currency by passing an instrument proficiency check in the category of aircraft involved. The check may be given by an FAA inspector, an FAA-approved check pilot, or an authorized instructor.

4) According to 14 CFR §61.133, commercial pilots must hold an instrument rating in order to carry passengers for hire on cross-country flights of more than 50 nautical miles, or to carry passengers for hire at night.

5) An airplane used for IFR flight must be equipped with the following instruments in addition to those required for VFR day and night flight:
   a) Two way radio communications system and navigational equipment appropriate to the ground facilities to be used; if a transponder is required, it must have been inspected and found to comply with FAA standards within the preceding 24 months.
   b) A gyroscopic rate-of-turn indicator.
   c) A slip-skid indicator.
   d) A sensitive altimeter adjustable for barometric pressure, The altimeter and static pressure system must have been inspected and approved for IFR flight within the past 24 months.
   e) A clock displaying hours, minutes, and seconds with a sweep-second pointer or digital presentation.
   f) A generator or alternator of adequate capacity.
   g) A gyroscopic pitch and bank indicator (artificial horizon).
   h) A gyroscopic direction indicator (directional gyro or equivalent).

6) According to 14 CFR §91.215 (b), an aircraft must be equipped with an operable 4096 code transponder having altitude reporting capability (Mode C) in order to be operated in the following airspace:
   a) Class A airspace.
   b) In Class B airspace and within 30 N.M. of a primary Class B airport.
   c) In and above Class C airspace.
   d) At and above 10,000 feet MSL excluding the airspace at and below 2,500 feet AGL.

7) A pilot must have a photo identification in his physical possession or readily accessible in the aircraft when exercising the privileges of a pilot certificate. The photo identification must be one of the following:
   a) Valid U.S. driver's license.
   b) U.S. issued federal or state identification card.
   c) U.S. Armed Forces' identification card.
   d) Official passport.
   e) Credential that authorizes unescorted access to a security identification display area at an airport regulated under 49 CFR part 1542.
   f) Other form of identification that the Administrator finds acceptable.

NOTAMs

1) A NOTAM, or Notice to Airmen, contains time-critical aeronautical information that is of either a temporary nature or not sufficiently known in advance to permit publication on aeronautical charts or in other operational publications. It receives immediate dissemination via the National NOTAM System.
2) FDC NOTAMs deal with regulatory information, including amendments to approach procedures, or changes in current aeronautical charts. They are available from FSS.
3) L NOTAMs, or Local NOTAMs, contain information such as taxiway closures, airport rotating beacon outages, and other information that does not affect instrument approach criteria. L NOTAMs are disseminated in local areas only.
4) D NOTAMs, or Distant NOTAMs, are disseminated for all navigational facilities. They are distributed automatically by appending to the hourly weather reports. They are available at air traffic facilities and FSS. Indefinite duration D NOTAMs are published in Class II NOTAMs Publication.

Two-Way Radio Communications Failure

1) UNDER VFR CONDITIONS -- If radio communications failure occurs under VFR conditions, or if VFR conditions are encountered after the failure, the pilot should continue the flight under VFR and land as soon as practicable.
2) UNDER IFR CONDITIONS -- If radio communications failure occurs under IFR conditions, the pilot should continue the flight according to the following:
   a) Route:
      i) The last route assigned by an ATC clearance; or
      ii) If being radar vectored, by the direct route from the point of radio failure to the fix, route, or airway specified in the vector clearance; or
      iii) In the absence of an assigned route, by the route that ATC has advised may be expected in a further clearance; or
      iv) In the absence of an assigned route or a route that ATC has advised may be expected in a further clearance, by the route filed in the flight plan.
   b) Altitude:
      i) At the highest of the following altitudes or flight levels for the route segment being flown:
      ii) The altitude or flight level assigned in the last ATC clearance received;
      iii) The minimum altitude (converted, if appropriate, to minimum flight level as prescribed in 14 CFR §91.121(c)) for IFR operations; or
      iv) The altitude or flight level ATC has advised may be expected in a further clearance.
   c) Leave Clearance Limit:
      i) When the clearance limit is a fix from which an approach begins, commence descent or descent and approach as close as possible to the expect further clearance time if one has been received, or if one has not been received, as close as possible to the estimated time of arrival as calculated from the filed or amended (with ATC) estimated time en route.
      ii) If the clearance limit is not a fix from which an approach begins, leave the clearance limit at the expect further clearance time if one has been received, or if none has been received, upon arrival over the clearance limit, and proceed to a fix from which an approach begins and commence descent or descent and approach as close as possible to the estimated time of arrival as calculated from the filed or amended (with ATC) estimated time en route.
3) TRANSPONDER OPERATION:
   a) Squawk Code 7600

Contact and Visual Approaches

1) **Contact Approach** -- If an aircraft on an IFR flight plan has at least 1 mile visibility, is clear of clouds, and can reach the destination airport under those conditions, a contact approach may be requested. This approach allows the aircraft to deviate from the published approach procedure and proceed to the destination airport by ground reference. A contact approach must be specifically requested by the pilot and authorized by ATC.

2) **Visual Approach** -- The visual approach authorization is initiated by ATC to expedite incoming IFR traffic, and is either accepted or rejected by the pilot. This approach also allows the aircraft to deviate from the published approach procedure, and therefore the clearance cannot be issued by ATC unless the aircraft can approach and land in VFR conditions.

Microbursts

1) Microbursts are small, intense downdrafts of air. The downdrafts produce damaging winds as the air spreads outward in all directions from the downdraft center upon reaching the surface. Although they are normally associated with mature thunderstorm activity, microbursts can occur in weaker, convective clouds that have little or no precipitation reaching the ground.

2) Due to the effect of the ground upon the downdrafts, microbursts contain both vertical and horizontal shear in a very small area. Since they occur near the ground, they are extremely hazardous.

3) The downdraft is normally less than 1 mile in diameter, and can be as strong as 6,000 feet per minute. At about 1,000-3,000 feet AGL, the air transitions to a horizontal outflow. This outflow can be as great as 2 1/2 miles in diameter.

4) As an airplane enters a microburst, it can have a headwind as great as 45 knots, which will turn to a tailwind of 45 knots after the center of the downdraft is passed. Therefore, the shear can be as much as 90 knots.

5) A microburst will tend to be short in duration, and will seldom last longer than 15 minutes from the time it strikes the surface to dissipation. It is not uncommon, however, for microburst activity to continue for as long as an hour, since multiple microburst activity in an area is not unusual.

Weather Data

1) The Forecast Winds and Temperatures Aloft Chart is prepared for eight different levels, and are available daily as 12-hour progs valid at 1200Z and 0000Z. The levels below 18,000 feet are true altitudes, while levels above 18,000 feet are pressure altitudes. Temperatures are given in Celsius, and arrows with pennants and barbs, as shown below, give wind direction and speed. The arrow points in the general direction of the wind, while the number by the end of the arrow denotes the middle digit of wind direction. The triangular pennant denotes 50 knots of wind, and each barb is 10 knots. One half barb is 5 knots.

Examples of Plotted Temperature and Wind:

0 degrees Celsius, wind 250 degrees at 15 knots

-11 degrees Celsius, wind calm or light and variable

-47 degrees Celsius, wind 360 degrees at 115 knots

-9 degrees Celsius, wind 260 degrees at 50 knots

2) The Low Level Significant Weather Prog Chart consists of four panels. The two lower panels are 12- and 24-hour surface progs. The two upper panels are 12- and 24-hour progs of significant weather from the surface to 24,000 feet. Conditions shown are forecast for valid time of chart. Low-level significant weather prog charts are issued four times daily.

3) The Observed Winds Aloft Chart is issued twice per day, and contain four panels of winds at different altitudes. The wind direction and speed are denoted the same as the forecast winds aloft chart.

4) The Convective Outlook (AC) describes the prospects for general and severe thunderstorm activity. There are three forecasts: Day 1 Convective Outlook (first 24 hours), Day 2 Convective Outlook (next 24 hours), and Day 3 Convective Outlook (the 24 hours after the Day 2 period). These forecasts describe areas in which there is a slight, moderate, or high risk of severe thunderstorms, as well as areas of general (non-severe) thunderstorms. The times of issuance for Day 1 are 0600Z, 1300Z, 1630Z, 2000Z, and 0100Z. The initial Day 2 issuance is at 0830Z and updated at 1730Z. The Day 3 is issued once per day.

Radio Magnetic Indicator (RMI)

1) The Radio Magnetic Indicator consists of a rotating compass card, a single-barred bearing indicator, and a double-barred bearing indicator. Each bearing indicator can be switched to display information on either a VOR or ADF station. The compass card, actuated by the aircraft's compass system, rotates as the aircraft turns. The magnetic heading of the aircraft is always directly under the index at the top of the instrument. The pointers indicate the course to the station. Radials from the station are read at the tail of the pointer.

Automated Terminal Information Service (ATIS)

1) ATIS is the continuous broadcast of recorded non-control information in selected high activity terminal areas. Its purpose is to improve controller effectiveness and to relieve frequency congestion by automating the repetitive transmission of essential but routine information. ATIS broadcasts are updated upon receipt of any official weather, regardless of content change and reported values. A new recording will also be made when there is a change in other pertinent data such as runway change, instrument approach in use, etc.

Partial Panel

1) A vacuum or pressure pump failure will cause the attitude and heading indicators to give incorrect information. The attitude indicator will show pump failure first, and will eventually end up showing nose down banked attitude. Upon failure, these instruments should be covered to avoid distraction.

2) It is imperative to include all direct and indirect indicators in the instrument scan, since a disagreement between instruments is an indication of instrument failure. Also, upon failure, the indirect instruments must be used to control the airplane.

3) When blocking of the pitot system occurs, two situations can develop:
   a) If the ram air input to the pitot head is blocked, the indicated airspeed may drop to zero.
   b) If the ram air input and the drain hole are blocked, the pressure is trapped in the system and the airspeed indicator will react as an altimeter. During level flight, airspeed indication will not change; during climb, airspeed indication will increase; and during descent, airspeed indication will decrease.

4) In many unpressurized aircraft equipped with a pitot-static tube, an alternate source of static pressure is provided for emergency use. If the alternate source is vented inside the airplane, where static pressure is usually lower than outside static pressure, selection of the alternate source may result in the following instrument indications:
   a) altimeter indicating higher than actual;
   b) indicated airspeed greater than actual;
   c) vertical speed indicating a momentary climb.

Attitude Indicator Errors

1) A normal turn causes precession of the gyro toward the inside of the turn. After roll out, the miniature aircraft will indicate a slight nose up attitude and bank in the opposite direction. The precession error is normally between 3° and 5°, and is corrected by the erecting mechanism.

2) A skidding turn precesses the gyro toward the inside of the turn. After return of the aircraft to straight-and-level, coordinated flight, the miniature aircraft shows a bank in the direction opposite the skid.

3) During a normal turn, acceleration causes an indication of a climb, while deceleration induces a precession error indicating a nose-down attitude.

Volume 7 – Questions 2.3 & 2.4

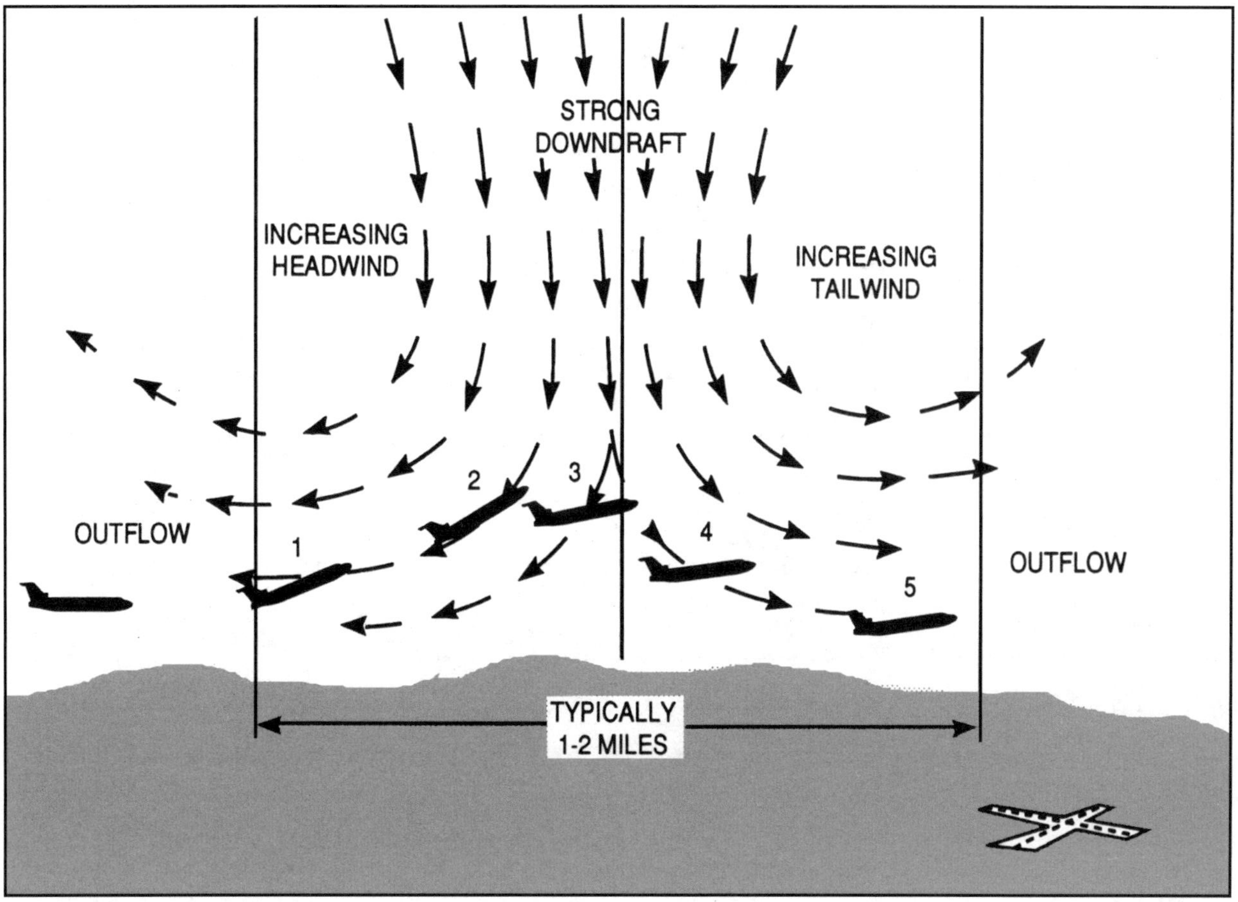

Volume 7 – Question 4.2

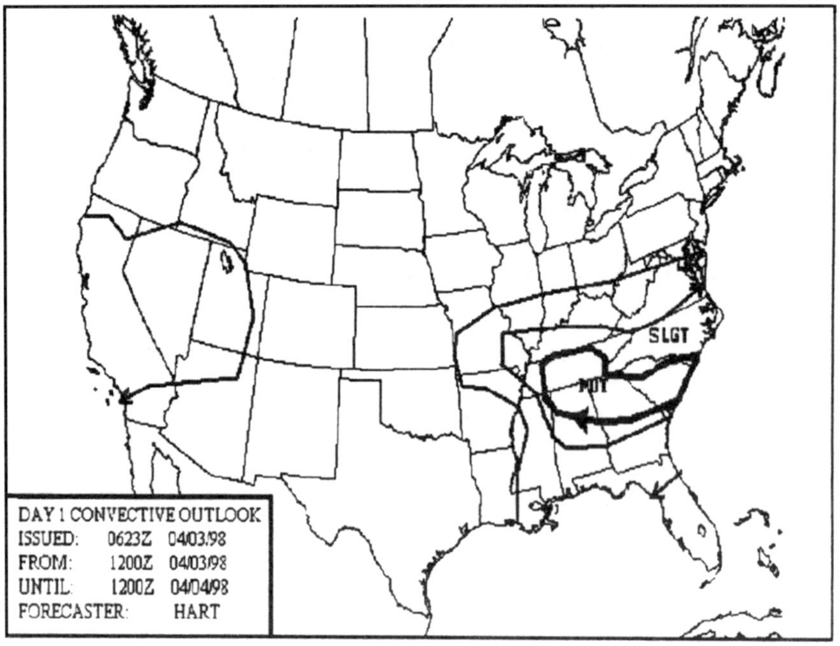

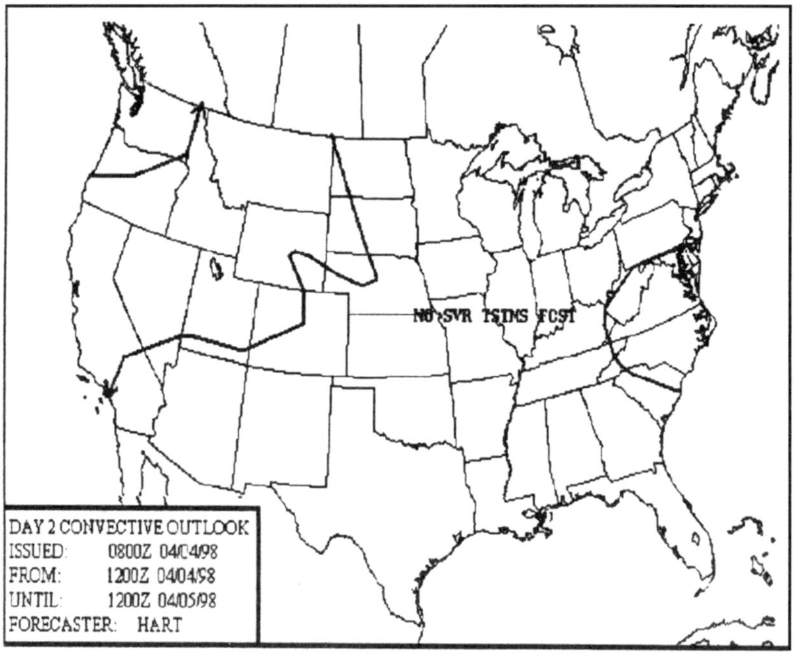

High Level Significant Weather Prognostic Chart – Volume 7 – Question 4.7

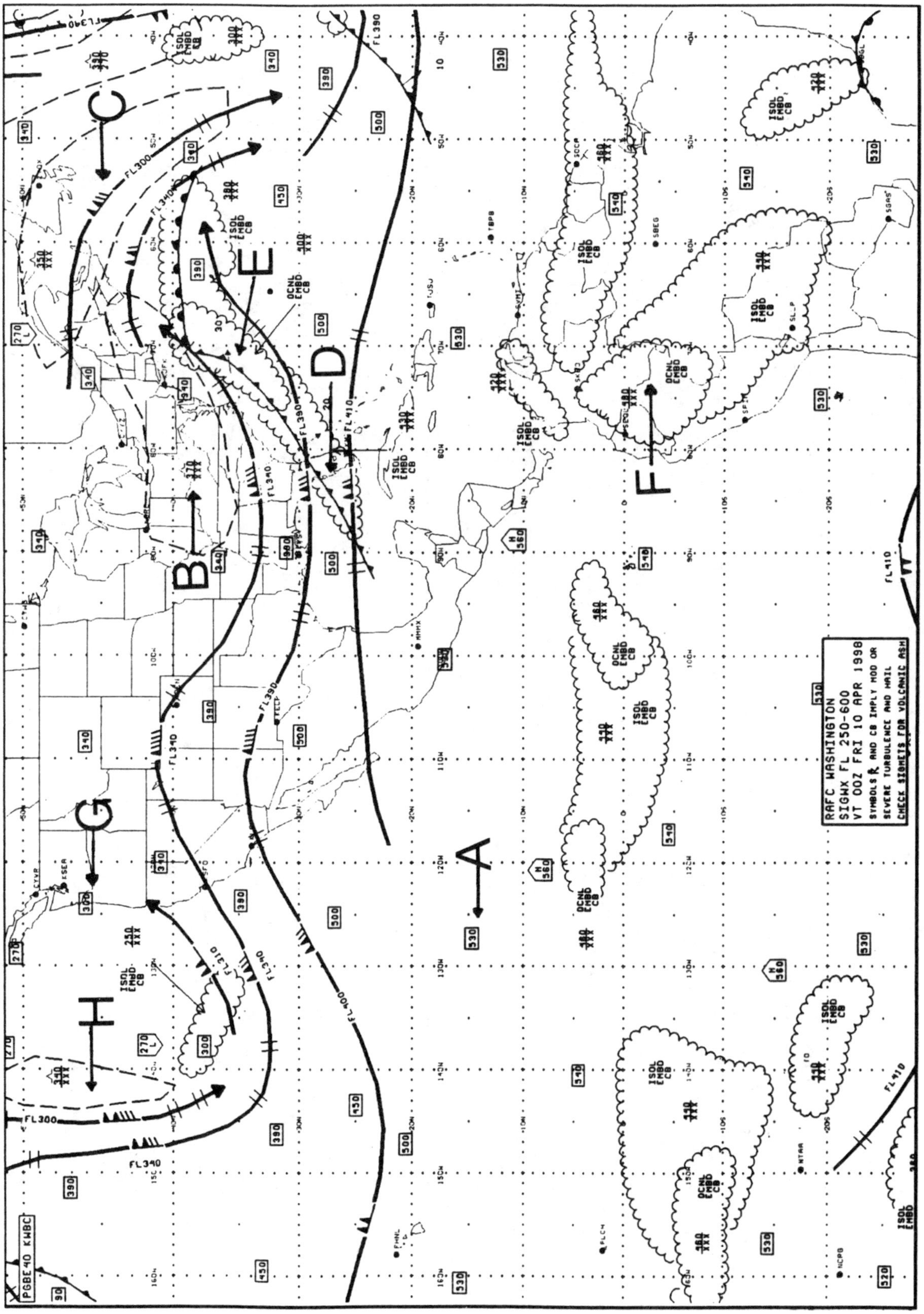

Low-Level Significant Weather Prognostic Chart – Volume 7 – Question 4.8

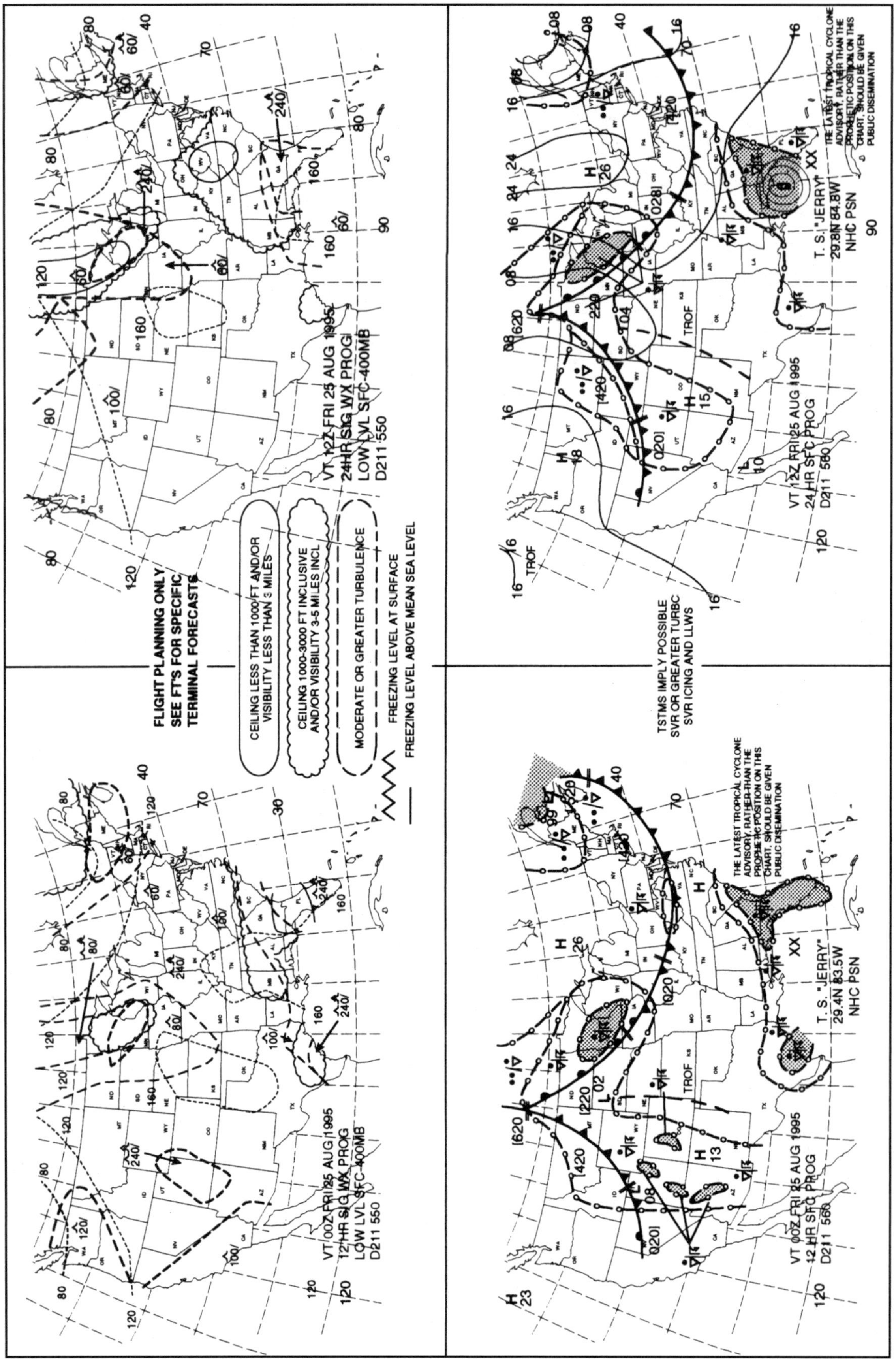

Volume 7 – Question 8.2

(VECTOR) (WASH2.WAGGE) 91094

WASHOE TWO DEPARTURE SL-346 (FAA)

RENO CANNON INTL
RENO, NEVADA

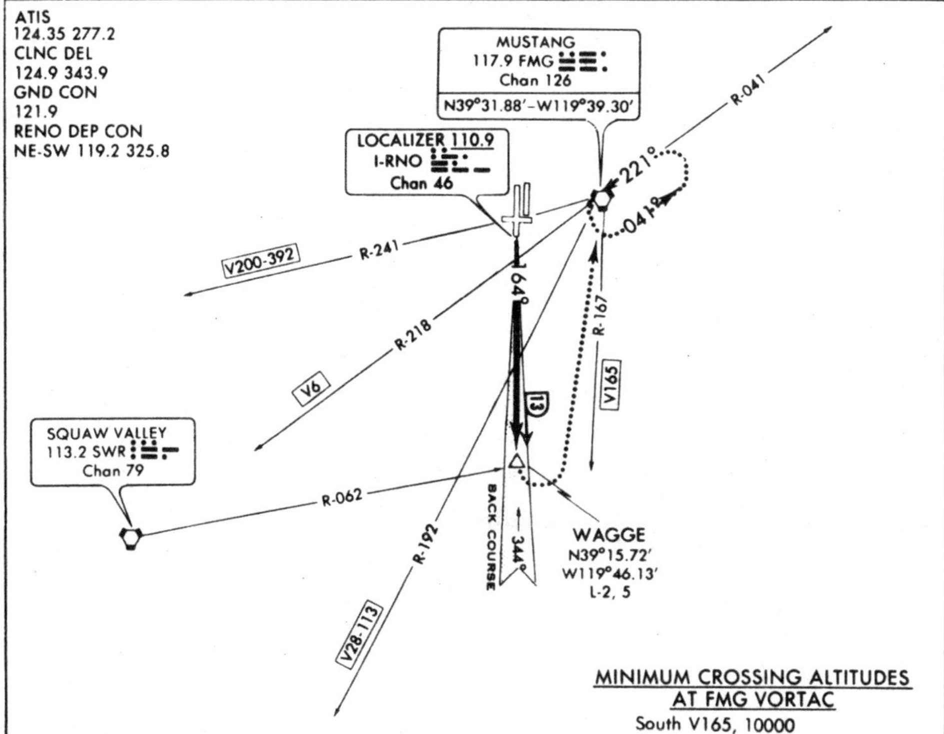

MINIMUM CROSSING ALTITUDES AT FMG VORTAC

South V165, 10000
Southwest V28-113, 10500
Southwest V200-392, 10000
Southwest V6, 12000

NOTE: Minimum climb rate at 400' per NM to 9000' required.

NOTE: Chart not to scale.

DEPARTURE ROUTE DESCRIPTION

TAKE-OFF RUNWAYS 16L/R: Climb via I-RNO Localizer south course to WAGGE INT then via radar vectors to assigned route.

LOST COMMUNICATIONS: If not in contact with departure control within one minute after takeoff, or if communications are lost before reaching 9000', continue climb via I-RNO localizer south course to WAGGE INT, turn left, proceed direct FMG VORTAC. Cross FMG VORTAC at or above MCA, thence via assigned route or climb in holding pattern northeast on FMG R-041, left turns to cross FMG VORTAC at or above MCA for assigned route.

WASHOE TWO DEPARTURE
(VECTOR) (WASH2.WAGGE)

RENO, NEVADA
RENO CANNON INTL

(Intentionally Left Blank)

Section 3 - Appendices and Supplemental Material

Appendix A – Composite Moisture Stability Chart

1) The composite moisture stability chart is a chart composed of four panels that depict stability, precipitable water, freezing level, and average relative humidity conditions. This computer-generated chart contains information obtained from upper-air observation data and is available twice daily with valid times of 00Z and 12Z.

2) The stability panel is the upper left panel of the composite moisture stability chart. This panel contains two indexes that characterize the moisture and stability of the air. These indexes are the K index (KI) and the lifted index (LI).
   a) The K index (KI) provides moisture and stability information. KI values range from high positive values to low negative values.
      i) A high positive KI implies moist and unstable air.
      ii) A low or negative KI implies dry and stable air.
      iii) KIs are considered high when values are at and above +20 and low when values are less than +20.
   b) The lifted index (LI) is a common measure of atmospheric stability. It is based upon the conditions of a hypothetical parcel of air being lifted to the 500-Mb level (approximately 18,000' MSL). LI values can be positive, negative, or zero. The LI does not identify the parcel's stability behavior at any of the intermediate altitudes between the surface and 500-Mb.
      i) A positive LI means a lifted surface parcel of air is stable and would resist upward motion. The more positive the LI, the more stable the air. Large positive values (+8 or greater) would indicate very stable air.
      ii) A negative LI means a lifted surface parcel of air is unstable would continue to rise and possibly gain increasing upward speed until stabilizing at some higher altitude. Large negative values (-6 or less) would indicate very unstable air.
      iii) A zero LI means a lifted surface parcel of air is neutrally stable has a tendency to neither rise nor sink. A neutrally stable parcel offers no resistance to vertical motion and, without further influence, would remain at the displaced level.
      iv) Temperature and moisture changes in the atmosphere change lifted index values.
   c) The KI and LI can be used in combination to assess moisture and stability properties of air masses.

3) The freezing level panel is the lower left panel of the composite moisture stability chart. This panel is an analysis of observed freezing levels. The freezing level is the height above MSL at which the temperature is zero degrees Celsius. The chart shows these values in hundreds of feet. BF is shown if the temperature is below freezing at the surface.

Composite Moisture Stability Chart

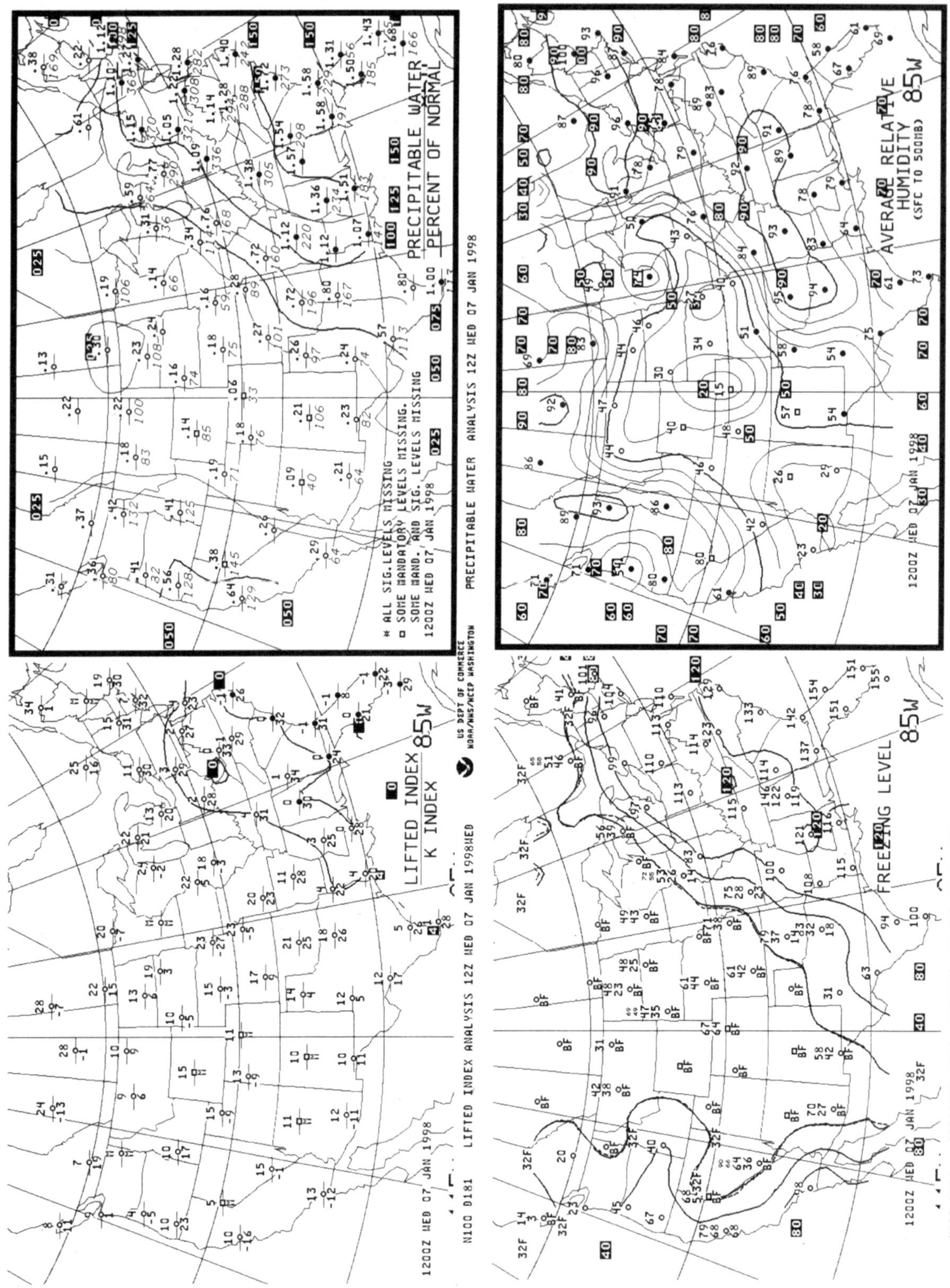

Appendix B – More about the Instrument Systems Preflight Procedures

1) While described in the PTS as an Instrument Cockpit Check, this check actually encompasses a number of checklist items that should be performed as a part of your overall preflight process. In the Instrument Flying Handbook (FAA-H-8083-15), the steps of this process are called the Instrument Systems Preflight Procedures.

2) Before any flight involving aircraft control by reference to instruments, you should check all instruments and their sources of power for proper operation. Refer to the AFM/POH of your airplane for detailed information regarding the checks noted below.

3) Before engine start:
   a) Review the aircraft records and confirm that the altimeter and static system, the transponder, and the altitude reporting system have been checked and approved within the past 24-calendar months. Ensure that the ELT batteries and VOR checks are up to date, as applicable.
   b) During the walk around inspection check the condition of all antennas and check the pitot tube and static port(s) for any obstructions. Ensure that all covers are removed and that there is nothing near the ports that could disrupt the airflow.
   c) Check the A/FD and all NOTAMs for the condition and frequencies of all navigation aids to be used during the flight. Current en route and approach charts for the departure, en route, destination, and alternate airports should be organized along with your flight computer and navigation log.
   d) Ensure that all radio equipment is turned off.
   e) Verify that the suction gauge has the appropriate reading.
   f) Ensure that the airspeed indicator has the proper reading. Typically, this should be zero while parked unless there is a significant headwind component.
   g) Set the altimeter to the current altimeter setting and check that the pointers indicate the elevation of the airport. Watch out for airports with multiple elevations.
   h) Check that the vertical speed indicator has a zero indication.
   i) Ensure that the heading indicator is uncaged, if applicable.
   j) Ensure that the miniature airplane of the turn coordinator is level and that the ball is approximately centered (level terrain).
   k) Verify that the magnetic compass is full of fluid and that the correction card is in place.
   l) Set the correct time on the clock.
   m) Ensure that the engine instruments have the proper readings.
   n) Verify that deicing and anti-icing equipment are available and have the appropriate fluid levels, if applicable.
   o) Ensure that the alternate static source valve can be opened and is fully closed.
   p) Turn on the master switch and listen to the electric gyro(s) as they spin up. Any hesitation or unusual noises should be investigated before flight.

4) After engine start:
   a) Turn the pitot tube heater on momentarily and watch for an ammeter deflection while turning it on. Use the method in the AFM/POH, if specified.
   b) Check the power source for the gyro instruments. The suction gauge should indicate the appropriate readings and/or the generator/alternator(s) should be operating appropriately.
   c) Check that the compass card moves freely and is accurate. You can check the indication, adjusted for deviation, against the known heading of a runway or parallel taxiway.
   d) Set the heading indicator to the compass heading after the gyro spins up (may be up to 5 minutes).
   e) Within approximately 5 minutes, the horizon bar on the attitude indicator should erect to the horizontal position and remain at the correct position for the attitude of the airplane. Any vibration should subside.
   f) Ensure that the altitude indicated on the altimeter is within 75 feet of the airport elevation when the set to the current altimeter setting. If off by more than 75 feet, recheck the instrument in the runup area using the TDZE for the runway as the basis for comparison to verify that the error is not related to variations in field elevation. If still off by more than 75 feet, the accuracy is questionable and the problem should be referred to a repair station.

g) Check that the vertical speed indicator has a zero indication. If not, have it adjusted or use the ground indication as zero while in flight.

h) Ensure that the carburetor heat is functional and in the off position, if applicable.

i) Verify that the radio equipment is operating properly and set as desired.

   i) Communication radios can be checked during calls to clearance delivery, ground, and tower at controlled fields and should be checked with the UNICOM at other airports.

   ii) VORs may be checked on the ground at some locations.

   iii) An ADF may be checked against a nearby NDB or radio station while on the ground if the weather permits reception of a signal.

   iv) The Localizer portion of the ILS may be checked on the ground if the airport has an ILS or other type of Localizer transmitter on the field. The instrument will show the side of the course on which you are currently located.

   v) A GPS should pass its self-checks, have an appropriate indication, and show the correct location. The database should be up to date if being used for approaches. Waypoints used for navigation should be verified as correct if the database is not up to date.

   vi) A LORAN should also pass its self-checks, have an appropriate indication, show the correct location, and the database should be up to date.

j) Ensure that deicing and anti-icing equipment is operational.

5) During taxi and takeoff:

   a) The turn coordinator should show proper turn indications during taxi turns and be level while straight. The ball should move freely and in the direction opposite the turn.

   b) The heading indicator should turn during taxi turns with minimal precession. Recheck the heading before takeoff and ensure that it agrees with the runway heading (within 5°).

   c) Ensure that the attitude indicator remains level during straight taxiing and does not tip more than 5° during turns. If not, it should be considered unreliable. Adjust the miniature airplane to a level flight attitude for the type of airplane and your seat position.

6) During and after engine shut down:

   a) Note any abnormal instrument indications.

   b) Listen for grinding or other unusual noises as the gyros spin down.

7) Postflight:

   a) Recheck the antennas and pitot-static sources while tying down the airplane.

   b) Document the failure or malfunction of any equipment or part of the aircraft and ensure that it is fixed as required (or that the appropriate parties are notified).

8) The Instrument Systems Preflight (and postflight) Procedures should be thorough but with practice and the development of a smooth flow, they shouldn't take up too much extra time. When in doubt about the condition of an instrument, it is best to err on the side of caution and stay on the ground until it can be fixed or until VFR conditions prevail.

Appendix C – More about Holding

1) When ATC requires you to hold, they will assign you a fix as a part of your holding clearance. This fix may be a navaid, an intersection, a DME distance from a navaid, or some other waypoint that you are capable of identifying based upon the equipment in your aircraft.

2) Upon arrival at the fix, an appropriate entry, which keeps your aircraft within the protected airspace for the holding pattern, should be made promptly.

3) When holding in wind, a symmetrical racetrack pattern will likely not be possible.
   a) You should compensate for the known wind except when turning.
   b) Adjust the timing of the outbound leg to achieve the required time for the inbound leg.
      i) Standard inbound leg at and below 14,000 feet – 1 minute.
      ii) Standard inbound leg above 14,000 feet – 1.5 minute.
      iii) Other time values may be assigned by ATC.
4) Holding legs may be the standard time based procedures or they may be DME based.
   a) Basic entry and holding procedures for DME holding are the same as time based holding.
   b) The DME distance for the *outbound* leg is given as part of the holding clearance.
   c) Use the DME distance readout as the end of the outbound leg.

Appendix D – Airspeed Changes and Rate Climbs & Descents

1) As indicated in Volume 1 of the DVDs, you should develop a knowledge of the pitch, power, and configuration combinations for the airplane you are flying that will produce various level flight airspeeds, climbs at various airspeeds and rates, and descents at various airspeeds and rates. You can create an index card with this information to act as a reminder in flight.
2) Changing your airspeed while in straight-and-level and turning flight will typically be accomplished through changes in the power setting. Introduction of drag producing devices may also be used in some aircraft and configurations but we will concentrate on the power setting.
   a) Adjust the power to the setting that you have previously determined will produce the desired airspeed.
   b) Adjust the pitch attitude to maintain altitude as the airspeed changes.
   c) Utilize the airspeed indicator as your primary power instrument and the altimeter as your primary pitch instrument as the speed is changing.
   d) Fine tune your power and pitch settings as the airplane stabilizes at the new airspeed.
   e) Adjust the trim to relieve the control pressures.
3) Constant rate climbs are entered in a manner similar to constant-airspeed climbs.
   a) Increase the power to the approximate setting required for the desired rate of climb.
   b) While increasing the power, raise the nose to the approximate pitch attitude required for the desired rate of climb.
   c) During the transition to the climb, use the airspeed indicator as your primary pitch instrument.
   d) Once the vertical speed indicator stabilizes, use this as your primary pitch instrument and the airspeed indicator as your primary power instrument.
   e) Adjust the trim to relieve the control pressures.
   f) Lead your level-off by 10% of your climb rate and adjust the pitch and power to the appropriates settings for your desired straight-and-level airspeed.
4) Constant rate descents are entered in a manner similar to constant-airspeed descents.
   a) Adjust the pitch and power to slow the airplane to the approximate descent airspeed desired while maintaining straight-and-level flight.
   b) Decrease the power further to the approximate setting required for the desired rate of descent.
   c) While decreasing the power, lower the nose to the approximate pitch attitude required for the desired rate of descent.
   d) Once the vertical speed indicator stabilizes, use this as your primary pitch instrument and the airspeed indicator as your primary power instrument.
   e) Adjust the trim to relieve the control pressures.
   f) Lead your level-off and adjust the pitch and power to the appropriates settings for your desired straight-and-level airspeed.

Appendix E – Instructor Certification for Instrument Rating Knowledge Test

NOTE: The endorsement below is representative of that required by 14 CFR §61.35 and §61.65(a)(4) and **MUST** be made in the applicant's logbook.

INSTRUCTOR CERTIFICATION

INSTRUMENT RATING KNOWLEDGE TEST

I certify I have reviewed the home study curriculum of (First name, MI, Last name) on the required training of §61.65(b). I have determined he/she is prepared for the Instrument Rating knowledge test.

Date: ______________________________

Signed: ______________________________

Certificate #: ______________________________

Expires: ______________________________